T0284830

HOW THE WORLD RAN OUT OF EVERYTHING

HOW THE WORLD RAN OUT OF EVERYTHING

INSIDE THE GLOBAL SUPPLY CHAIN

PETER S. GOODMAN

MARINER BOOKS

New York   Boston

HarperCollins books may be purchased for educational, business, or
sales promotional use. For information, please email the Special Markets Department at SPsales@harpercollins.com.

FIRST EDITION

Library of Congress Cataloging-in-Publication Data has been applied for.

ISBN 978-0-06-325792-4

24 25 26 27 28 LBC 5 4 3 2 1

For Deanna, my never-ending supply chain of
love, brilliance, and grace

CONTENTS

Part III: Globalization Comes Home

"THE WORLD HAS FALLEN APART."

Southern California appeared to be under siege from a blockade.

More than fifty enormous vessels bobbed in the frigid waters of the Pacific Ocean, marooned off the twin ports of Los Angeles and Long Beach. As days stretched into weeks, they waited their turn to pull up to the docks and disgorge their cargo. Rubberneckers flocked to the water's edge with binoculars, trying in vain to count the ships that stretched to the inky horizon.

This was no act of war. Rather, this was what it looked like when the global economy came shuddering to a halt.

It was October 2021, and the planet was seized by the worst pandemic in a century. International commerce was rife with bewildering dysfunction. Basic geography itself seemed reconfigured, as if the oceans had stretched wider, adding to the distance separating the factories of China from the superstores of the United States.

Given the scale of container ships—the largest were longer than four times the height of the Statue of Liberty—any single vessel held at anchor indicated that an enormous volume of orders was not reaching its intended destination. The decks of the hulking ships were stacked to the skies with containers loaded with seemingly every conceivable component of contemporary life—clothing, electronics, auto parts, furniture, refrigerated fruits, toys, medical equipment, bottled beverages, and drums full of chemicals used to concoct other products, from paint to pharmaceuticals.

Among the ships held in the queue was the *CSCL Spring,* a Hong Kong–flagged vessel that was carrying a whopping 138 containers from Yihai Kerry International, a major Chinese agricultural conglomerate. Together, they held 7.3 million pounds of canola meal pellets—enough animal feed to sustain twenty thousand cows for a week. This was exacerbating shortages of feed afflicting American livestock producers.

Five ships in this waylaid flotilla were collectively hauling 13 million pounds of Fiji bottled water. More than 17 million pounds of Heineken beer was held up. The Singapore-flagged *Wan Hai 625* was carrying almost 3 million pounds of polyethylene terephthalate resin, a key element for manufacturing synthetic fabrics and plastic bottles used to package soft drinks—another commodity in short supply. The same ship held 5.2 million pounds of solar panels and 1.6 million pounds of material for chain-link fencing. The Cyprus-flagged *Hyundai Hongkong,* just in from the South Korean port of Busan, held 64,511 pounds of vehicle carpet fabric that was destined for Tesla vehicles along with 28,723 pounds of cornhole game equipment.

By one estimate, the ships waiting off Southern California's two largest ports were collectively loaded with more than $25 billion worth of goods. And this was a mere fraction of the wares stranded by a global breakdown that had reached staggering proportions. Nearly 13 percent of the world's container shipping fleet was floating off ports from China to North America to Europe. Upward of $1 trillion worth of product was caught in the congestion.

All of this stuff was supposed to be somewhere else.

But the docks were overwhelmed by an unprecedented influx of containers as Americans stuck in quarantine outfitted themselves for the apocalypse, filling their basements with exercise bikes, their bedrooms with office furniture, and their kitchens with baking equipment. Most of these goods were manufactured in Asia. The trucking industry complained that it could not hire enough drivers to move this tsunami of product. Warehouses were stuffed to the rafters and short of workers. The railroads—hollowed out by years of corporate cost cutting—were buckling in the face of a surge of demand.

So tens of thousands of containers sat stacked at the ports, waiting for someone to haul them to their next destination. Out on the water, miles from shore, ships lay at anchor, their listless crews wondering when they would next encounter land.

Here was the central explanation for why Americans suddenly found themselves unable to buy everything from medical devices and hand sanitizer to toothpaste and smartphones. The scene off Southern California was the reason that carpenters could not find wood, why families painting homes were settling for whatever color was available, and why hospitals were substituting subpar medicines for the ones they could not procure.

For decades, the world had seemed compressed and tamed, the continents bridged by container ships, internet links, and exuberant faith in globalization. Now, the earth again felt vast and full of mystery.

In the center of the pileup off Long Beach lay the *Maersk Emden,* a Danish-flagged container ship that stretched 1,200 feet long and 158 feet wide, making it more than five times the size of New York's Grand Central Station. Freshly arrived from the Chinese port of Ningbo, it was carrying roughly twelve thousand containers.

That haul included 474 containers for LG, the South Korean home appliance brand, and 74 boxes full of Nike products. Mattel and Hasbro, two giant toy companies, were collectively waiting for 160 boxes aboard the *Emden.* And just as the weather was turning cold in North America, 48 containers shipped by Burlington Coat Factory were stranded atop the ship.

Hagan Walker had only one box on board the *Maersk Emden,* a forty-foot container logged in the shipping manifest as MSMU8771295. But it held the most important order in the brief history of his start-up.

Walker's company, Glo, was based in a small town in Mississippi. It made plastic novelty cubes that lit up when plunked in water. He had recently secured a breakthrough deal—a contract to make bath toys for *Sesame Street,* including a figurine version of the iconic Elmo. He had planned to debut them during the pivotal holiday season, now only two months away.

Like millions of companies, Walker's operation depended on two key elements: factories in China to make its products, and gigantic container ships to carry them to American shores. For decades, this had proven a cheap and reliable way to do business, the means by which major brands and niche players alike had kept the world's largest economy stocked with everything from oven cleaner to aircraft parts.

But that equation was unraveling.

Walker's year had been dominated by logistical torment. Chinese factories were short of workers and raw materials, yielding alarming delays. The shipping industry was deluged. Trying to book cargo space on a container had become next to impossible. The cost of transporting goods across the Pacific had multiplied tenfold.

In the middle of it all, on March 23, 2021, like a farce engineered by the heavens, a behemoth container ship had gotten lodged inside the Suez Canal in Egypt. Traffic through the canal came to a stop, halting a huge volume of goods moving from Asia to Europe. For months afterward, factories around the globe waited for shipments of parts and raw materials, impeding their production and threatening the livelihoods of their workers.

Walker had managed to navigate this shifting assortment of threats. Yet now, in the fall of 2021, he found himself confronting the most frustrating ordeal yet—the mother of all traffic jams off the coast of Southern California.

As the calendar continued its relentless march toward the holiday season, Walker's Elmo dolls were floating out on the water, castaways during what I had come to call the Great Supply Chain Disruption.

For society as a whole, the stakes were considerably greater than whether a collection of light-up *Sesame Street* toys made it to their final destinations in time for Christmas. Still, Walker's container was an ideal object to follow—as we will do in the course of this book—because its passage traced the voyage of countless other loads of cargo traversing the Pacific and the American continent. Its odyssey, at once commonplace and astounding, afforded an ideal vantage point on the breadth of the troubles assailing the global supply chain. The journey of that single box revealed how and why trillions of dollars' worth of goods, some critical to saving lives, effectively went missing in the midst of a public health catastrophe.

By the time the *Maersk Emden* joined the floating queue off Long Beach bearing Walker's shipment, people from Europe to Africa to North and South America had endured a terrifying scarcity of personal protective gear like face masks and medical gowns. This had forced frontline medical workers to attend to patients stricken with COVID-19 absent adequate protection.

Society had experienced the disappearance of toilet paper from store shelves around the globe amid panicked hoarding. Women's sanitary products had become difficult to find, along with medicines like antibiotics and even aspirin. Meat display cases at supermarkets sat empty. For a time, Grape Nuts, the popular breakfast cereal, all but vanished, along with tapioca beads used to make boba tea.

Factories in Asia that manufactured computer chips could not keep pace with a dramatic increase in demand, an emergency in an age in which chips had become the brains for all manner of devices. Auto factories from Japan to the United States to Brazil halted production, citing a lack of chips. American car dealers typically held two to three times as many vehicles as they sold in a month. By the end of 2021, their inventory had plunged to a record low—less than half their volume of sales. And as new cars became scarce, even used vehicles saw their prices explode.

Medical device manufacturers embarked on a largely futile campaign to shame chip companies into prioritizing their orders over those from smartphone companies like Apple and Google. Major electronics companies began covertly buying old toys and video gaming consoles, breaking apart ancient PlayStations and Barbie accessories to harvest the chips within.

Even supplies of baby formula were exhausted, plunging millions of American families into a state of desperation.

For ordinary people who had never previously had reason to contemplate the intricacies of the global supply chain, all of this was cosmically disconcerting. The shortages of goods conveyed a gut-level affirmation that contemporary life itself had gone haywire, exposing a dark and unsettling truth: no one was in control.

In wealthy countries, contemporary society had been steeped in the idea that the internet had transcended the traditional constraints of time and space. You could go online at any hour, on any day, no matter the weather, click here, and then wait for the truck to arrive bearing your order.

In a world full of grave uncertainty, here was a sure thing.

The supply chain was not just the circulatory system for goods, but also the source of a deep-seated sense of authority over human circumstance, and a rare unifying aspect of modern existence. In a time of flagging faith in government, skepticism about the media, and suspicion of corporate motives, everyone could at least believe in the unseen forces that brought the UPS guy to their door. The links connecting farms, factories, and distribution centers to households and businesses had seemed inviolate.

We had no illusions about the moral sanctity of the bargain. Most of us understood that the businesses that dominated the supply chain were making the economy more unequal, enriching executives who frequently abused the rank and file, poisoning our democracies and sowing toxicity in our political discourse, to say nothing of the natural environment. To the extent to which we thought about it, we generally recognized that our mode of consumerism was threatening humanity with extinction via climate change, while exploiting labor from South Asia to Latin America.

We grasped that Amazon was run by a bazillionaire, Jeff Bezos, whose fortune was so vast that he could blast himself into space even as he failed to provide his warehouse workers with face masks in the midst of a deadly pandemic. But we also understood what we were buying: certainty and security, the comfort of not having to worry about running out of whatever we needed. In exchange for our tacit assent in the often-unsavory terms of global capitalism, we gained convenience and reliability to a degree that was unimaginable mere decades ago.

Which meant that the breakdown in the system was bigger than the delays and the shortages of goods. It forced us to contemplate the possibility that everything was spinning out of control, and at the worst possible moment.

As the supply chain began fraying, urban reality from Minneapolis to Milan was dominated by the ceaseless wailing of ambulances hauling those stricken with COVID-19 to hospitals, where people were dying on gurneys stashed in corridors, the rooms overflowing, the supply of ventilators exhausted. From San Francisco to Stockholm, people were taking their last breaths alone in nursing homes, without saying goodbye to their children and grandchildren. Every day brought grim reports of a rising tide of death that eventually took the lives of nearly 7 million people worldwide.

People were succumbing to existential dread that was testing faith in everything, from the wisdom of public health authorities to the enduring strength of their marriages. The supply chain failures added to the emotional strains of life under lockdown—the terror, the claustrophobia, the tedium. Denied access to supermarkets and restaurants, unable to send our children to school or interact with our friends, consumed with fear for ourselves and our loved ones, we were more dependent than ever on the delivery system for goods.

When even that failed, the emotional consequences cut deep.

My own family was based in London during the first two years of the pandemic, and well aware of our relative privilege amid the deprivations of lockdown. My wife, Deanna, and I were both able to work from home. We had outdoor space for our two older children, and ample internet bandwidth to accommodate their suddenly remote education. We had space to stockpile popcorn, toys, and jigsaw puzzles—anything that might provide a momentary diversion from the stress and angst. Still, we were consumed by a looming event of seismic proportions for our household: eight years after the extremely premature birth of our daughter, Deanna was due to give birth to our third child in early April 2020.

I watched my wife remain stoic as one plan after another succumbed to the wrenching realities of the pandemic. Her parents could not fly in to help with the baby, and no friends could stay with our older children during the birth. The peaceful parental leave we had envisioned devolved into conditions that felt like wartime. I watched her deliver our baby—a healthy, perfect boy—knowing that I would be forced to leave her side within minutes of meeting him, that our older children, who had for years begged for another sibling, would not be allowed to visit him in the hospital.

I watched Deanna accept the fact that our son's very existence could not be verified by any birth certificate, because government offices were closed. When the hospital discharged her within a day and a half after a medically necessary C-section, I helped her climb the staircase to our bedroom, enraptured, with the baby in her arms. I watched her nurse him, recording his feeding times, diaper output, and weight with the understanding that no pediatrician would be able to evaluate him in person.

Then Deanna tried and failed to order toilet paper, which I had resisted panic-buying. She tried to order hand sanitizer, then resourcefully sought to make it herself, only to discover that the key ingredients—rubbing alcohol and aloe gel—were out of stock everywhere. She tried to order disinfectant wipes and N-95 masks. She tried to order backup baby formula. It was only when she found herself unable to summon these necessities to our doorstep that I finally saw her break down.

Rationally, Deanna knew that our lives were not in imminent danger, that we were still among the most fortunate people on the planet. Yet on an emotional and practical level, life as we knew it had ended.

"All we thought to wish for was a healthy baby," she told me. "And now he's here, but the world has fallen apart."

Something foundational had indeed broken. A basic mechanism of contemporary life was suddenly inoperative, yielding the stupefying impossibility of buying simple things that had previously been everywhere.

What on earth was going on?

The answer, as I've come to understand it, was nothing less than the breakdown of globalization.

Over recent decades, multinational companies from North America to Europe to Japan had placed their fate in a ruthless sort of efficiency. They had steadily entrusted production to factories around the globe, and especially plants in China, chasing lower costs and fatter profits.

And they had behaved as if this strategy was devoid of risk, as if China's industrial parks might as well have been extensions of Ohio and Bavaria, because low-cost shipping was assumed to be an immutable reality. They either did not know or did not care that the shipping industry was basically a cartel, operating largely beyond the oversight of any government watchdog.

Once their products reached American shores, companies relied on transportation networks that depended on millions of workers who submitted to dangerous, lonely, and soul-crushing jobs, even as their pay and working conditions were downgraded to free up cash for shareholders. In constructing a supply chain governed by the relentless pursuit of efficiency, trucking and railroad businesses treated their workers as if their own time was both limitless and without value, deserving of no compensation for hours stuck waiting for the next load.

In a quarter century of writing about economics from Asia to Europe to North America, I've frequently been exposed to a conventional narrative in which the supply chain is typically discussed as separate from the rest of commercial existence—a distinct network of connection points linking factories to customers, spanning ships and trucks and warehouses. This is wrong. What we refer to as the supply chain is inextricable from the broader economy. How it operates—who gets paid, who bears risk, who is allowed to work at home during a pandemic, and who must stand in harm's way— reflects the same power dynamics and values that shape the rest of life.

From the railroads to trucking firms to warehouses, major companies had long treated their workers like costs to be contained rather than human beings with families, medical challenges, and other demands. Employers assumed that they did not have to worry about running out of laborers, even as they engaged in wanton exploitation. Unions were weak and working-class people were desperate for a paycheck—so much so that adequate numbers could presumably be counted on to submit to the grueling undertaking of delivering freight, processing meat, and other brutal yet critical functions of the supply chain.

At the same time, decades of zealous reverence for deregulation as the solution to nearly every problem served to cede economic fate to a handful of monopolistic companies that dominated key industries, from the railroads to meatpacking. Here was the outgrowth of a fundamental alteration of American capitalism over the last half century, the elevation of shareholder interests to primacy, the triumph of financial considerations over all others. The executives running corporations that controlled critical swaths of the economy—manufacturing, transportation, food processing—answered not to their customers or local communities but to distant yet omnipotent bosses: the money managers of Wall Street. They had cut capacity as a way to limit supply and boost the prices for their products and services. That had left markets prone to disarray and shortages whenever trouble arrived.

In Washington, both major political parties had long placed faith in the fantastical notion that gigantic companies left to seize commanding holds over their markets would yield greater efficiency—a concept that took on cultish currency. Far from happenstance, this represented a triumph of corporate lobbyists deployed over decades. Successive presidential administrations and Congress dismissed decades of lessons about the perils of unchecked monopoly power, deactivating basic antitrust law. They bought into the idea—or at least the accompanying campaign contributions—that scale secured through mergers was the best way to supply consumers with abundant choices and lower prices.

The pandemic exposed the dangers of all of these assumptions.

It laid bare the consequences of relying on faraway factories and container ships to keep humanity supplied with goods.

It exposed as reckless the world's heavy dependence on a single country—China—for critical products like protective gear and medicine, and especially as Washington and Beijing were locked in a trade war.

It revealed the risks of leaning on transportation systems staffed by people whose wages and working conditions had been decimated by cost cutting, leaving companies wide open to revolt.

And it demonstrated the pitfalls of handing responsibility to monopolists for the supply of meat, baby formula, and other basic fortifications—a recipe for skyrocketing prices and empty shelves.

Unregulated behemoths left to dominate markets in the name of efficiency turned out to yield results that were efficient only on Wall Street.

The United States, the world's most powerful country, had successfully mobilized to defeat the Nazis, put humans on the moon, and catalyze the computer age. Yet, in the face of a lethal pandemic, it could not make or secure enough face masks to outfit medical workers. It was unable to build enough ventilators to allow its most vulnerable citizens to count on their next breaths. And it could not secure adequate stocks of basic medicines. Here was a repudiation of the cult of efficiency that had long ruled American business.

Still, if COVID-19 produced the shock that was the most immediate cause of this pronounced supply chain distress, it merely unmasked vulnerabilities that had been there all along, mounting over decades.

Trucking had long depended on a steady stream of recruits to compensate for the fact that it was a horrifying way to make a living, one subject to the predatory financial shenanigans of employers. American railroads had frequently been looted by their investors and run less as transportation systems than as wellsprings of dividends—a story that went back to the Robber Barons of the nineteenth century. From meatpacking plants to warehouses, the people producing goods and delivering them to our homes had long been forced to choose between their lives and their livelihoods.

The pandemic did not create this situation, but it brought the consequences into stark relief.

And the broad chaos in the global supply chain helped deliver another economic affliction: inflation.

By early 2022, in the name of snuffing out price increases, central banks around the world would begin lifting interest rates. This would foist higher borrowing costs on homeowners and credit card holders. It would threaten ordinary workers with joblessness while depressing stock prices. Though economists debated the causes of inflation, part of the blame clearly fell on the reality that astonishing quantities of goods were stuck floating off ports. Distress was rippling around the globe, and ordinary people were paying the mounting costs.

By early 2023, the worst disruptions of the pandemic years had subsided. The floating traffic jams had all but disappeared, shipping rates had plunged, and product shortages had eased. Yet the same foundational perils remained, awaiting an inevitable future disturbance.

The global economy has entered a new era of enduring volatility. As climate change alters the natural realm, the global supply chain will be subject to new rules and constant reassessment of risks. Russia's assault on Ukraine has enhanced the prospect of the world splintering into rival camps, complicating the geography of international trade. China and the United States appear locked in a cold war whose consequences are playing out around the globe, reshaping alliances, trade pacts, and fundamental understandings about the nature of international engagement.

These are monumental variables. We do not know when the next shock will arrive or where it will occur, but we can be certain that such a moment is coming.

To meet the challenge of the next disturbance, we need to grapple with how we got here. We need to understand how the supply chain became so complex, extended, and centered on a single country. And we must reconfigure the supply chain to safeguard society through greater resilience.

The executives of publicly traded corporations and their hired enablers in the political sphere have played make-believe with the world economy. They have disregarded the enormity of the globe while willfully ignoring the dangers of a supply chain dependent on desperate workers and greedy monopolists. They have done so repeatedly, forsaking the lessons of previous disasters, because this has enriched investors while undermining the interests of everyone else—from workers to consumers to medical patients.

The globalization to which we have become accustomed was propelled by an especially intoxicating form of efficiency, a concept known as *Just in Time,* or *lean manufacturing.* Under that banner, multinational companies have skimped on the costs of stashing backup parts and products in warehouses. Instead, they have depended on the Web and container ships to summon what they need in real time.

Companies adopted this mode because spending less on warehouses freed up cash that they could lavish on shareholders. They went too far, damaging the vibrancy of the supply chain in exchange for the immediate gratification of investors.

The shortages of the pandemic have prompted some companies to recalibrate, building up inventories as they pivot from Just in Time to Just in Case.

As the United States and China treat one another like rival powers, multinational companies have shifted some factory production to other countries like Vietnam. American businesses are setting up factories in Mexico and Central America to retain low-cost manufacturing without having to contend with the vagaries of the Pacific Ocean. And some companies are embracing so-called reshoring, bringing factory production back to the United States.

All of these objectives are being accelerated by deepening concern over the increasingly palpable manifestations of climate change. As companies consider their carbon footprints—prodded by activists and the omnipresent possibility of losing public favor—they are inclined to manufacture their goods closer to their customers.

At investor gatherings and think-tank conferences, executives now pledge fealty to resilience and sustainability in the same way they used to celebrate efficiency. Yet the traditional underpinnings of the supply chain remain. Once the current crisis recedes, the old incentives are likely to reemerge.

Same as ever, an executive who cuts costs by moving production to a low-wage factory is likely to cash a bonus check. The same corporate chieftain who delivers on the rhetoric about resilience, spending money on warehouses instead of dividends, risks losing access to the corporate jet.

Which means the rest of us remain vulnerable to the impacts of the next reverberation.

This book unfolds in three parts. In part one, we will follow Hagan Walker's order from the Chinese factory that produced it to the port of Long Beach, California. Along the way, we will examine how China rose to become the center of global manufacturing. We will explore how multinational corporations employed a cynical narrative in adding China to the ranks of the World Trade Organization, casting this as a boon for freedom. And we will probe the key elements that shaped the era of China-centric globalization—the advent of container shipping and the embrace of Just in Time. We will examine how business consultants took a sensible idea and turned it into a crude imperative to slash inventory, setting the world up for crisis. We will explore the legacy of Henry Ford, a supply chain pioneer who feared this very disaster.

In part two, we will trace the next leg of the journey of Walker's container, from Southern California across the American continent to Mississippi. En route, we will probe the workings of every industry involved in its passage, from shipping and trucking to rail, while paying special attention to the fragility delivered by headlong deregulation. Through the experiences of slaughterhouse workers and cattle ranchers, we will explore how monopolists have engineered scarcity to boost their prices while profiting off calamity.

Part three takes the crisis as an opportunity to reinvent the supply chain. We will follow Walker in pursuit of an alternative to Chinese factories, and then examine the promise of reshoring—bringing production back to home markets. Finally, we will explore the growing importance of regional approaches to manufacturing, examining the investment flowing into Mexico from companies seeking closer access to American consumers.

At the end of our harrowing journey lies one singular truth: humanity has come to depend on a disorganized and rickety global supply chain for access to the products of our age, from lifesaving drugs and computer chips to toys and games. The system relies on myriad forms of labor exploitation, which has made it perpetually vulnerable to breakdown. And it has been constructed as a means of rewarding the investor class, often at the expense of reliability.

The Great Supply Chain Disruption is not some curious piece of recent history. It is a preview of the dysfunction that surely lies ahead if we fail to get the machine in order.

PART I

The Great Supply Chain Disruption

CHAPTER 1

"JUST GET THIS MADE IN CHINA."

The Origins of the Factory Floor to the World

Hagan Walker would have preferred to make his light-up cubes in his own country.

He was a kid from the rural American South who valued family ties. A proud graduate of Mississippi State University, Walker had kept his start-up company in his college town of Starkville, population twenty-five thousand, rather than chase the allure of the usual places for a gifted engineering graduate—Silicon Valley, Austin, New York.

He did not indulge the vernacular of the technology crowd, with their "stakeholder solutions," "giving back," and other terms that depicted their businesses as messianic missions. But he was rooted in Mississippi, a state that needed jobs more than most others. He felt like he was part of something bigger in staying close to home and running a business that was providing twenty-seven local paychecks.

Even so, the same forces that had already led millions of other American businesses across the Pacific eventually carried Walker there, too. In the early years of the twenty-first century, in the age of the container ship and the internet, China was typically the obvious place to make nearly everything.

It was not simply that factories in China offered the lowest prices, though this was frequently true. It was that all the components and raw materials were available in abundance, along with manufacturing know-how.

Over decades, the spread of production to China had catalyzed investment into the attendant industries—chemicals, plastics, glass-making, rubber processing, and the smelting of every conceivable type of metal. Ports, logistics companies, and shipping agents could deliver finished products wherever container vessels docked. Vast digital marketplaces connected companies around the globe with local factories. You could send over plans and specifications for a product and quickly receive back a sample. You could fine-tune the process with an exchange of emails or texts, then wire over some money, and soon unload a container full of finished goods.

By contrast, the United States increasingly felt like a place where making physical things was a forgotten art.

In going to China, Walker had followed the path of least resistance.

A tinkerer by nature, Walker had mastered a power screwdriver when he was only three, a fact that became evident to his parents when they discovered he had used it to remove the deadbolt from their front door.

As a high school student, he launched a computer repair business, earning enough to invest in mutual funds whose winnings provided the seed capital for Glo, the company he started in 2016, just after graduating from college.

His initial idea for the business centered on a novelty item that could punctuate cocktails: lumps of plastic shaped like ice cubes that lit up when inserted into water. They would help bartenders recognize when customers were ready for a refill—when the light turned off.

He and an early-stage partner pitched the idea for the cubes to a campus entrepreneurship center, securing a $1,000 grant that they used to buy a 3D printer. In Walker's senior year, he and his partner won an Entrepreneur Week competition, netting another $15,000 that they used to launch the business, purchasing injection molds and tooling—the gear needed to make the cubes.

Walker gained a coveted internship at Tesla and spent the summer of 2015 in California. He was offered the chance to stay on via a full-time job that would have paid him $130,000 a year when he was only twenty-four. But he was captivated by the prospect of running his own company. So he returned to Mississippi to start Glo.

He moved the company into a grand old movie theater just off Main Street that had been abandoned for decades. Next to the front door, he installed an old Coca-Cola vending machine—classic glass bottles and all—a marker of retro-reverence in a town whose forsaken brick warehouses were being revived as gastropubs and coffee shops. His employees brought their children and dogs to work.

Walker sold his first products to local bars. Then, he heard via Facebook from an unexpected customer, the mother of a four-year-old boy who was autistic. Bath time had been a perpetual nightmare. Her son suffered from sensory overload and was terrified by the sound of rushing water. When she deposited a Glo cube into the bath, the toy transfixed and soothed him.

Walker had by then teamed up with another recent Mississippi State graduate, Anna Barker, who seized on this anecdote as a revelation.

Barker had grown up in the southern reaches of Mississippi, in a small town short of possibilities and long on dysfunction. A high school teacher had once counseled her to stop wasting time making college plans and take the cheerleading scholarship offered by her local community college. Barker had ignored that advice. Now she recognized an opportunity. The Glo cubes were not just playful knickknacks; they were the potential building blocks for a media business centered on characters she would develop, using them as the basis for a new line of light-up figurines.

She sketched out conceptions, developing their personalities, accompanying stories, and animated features. The company would sell doll versions of her characters, with the cubes forming their glowing midsections. Much as Disney had cultivated an appetite for merchandise tied to characters encountered in its cartoons, Glo would sell its cubes as a complement to an imagined world.

At a toy industry trade fair in Dallas in October 2019, Barker met with a giant in the realm: *Sesame Street*. The iconic children's education program was planning to roll out a new autistic character, Julia. By early the following year, Glo had a contract to make not only the Julia figurines but also a version of Elmo.

Walker and Barker had secured the most important assignment in the brief history of their company. Now, they had to line up a factory that could make tens of thousands of their new *Sesame Street* characters and ship them to the center of Mississippi. And they had to get these new characters to their warehouse in time for the holiday season of 2021.

"We hadn't done something of that magnitude before," Walker said. "I tell people all the time, 'Either this company is going to go somewhere, or I'm going to be sleeping on your couch.'"

When Walker first began manufacturing his Glo cubes, he had tried to rely on American production. One American plant offered to make the steel plates used to manufacture the cubes for $18,000—twelve times the going price in China. Another domestic supplier offered to make the cubes for a little more than the China price, but it turned out to be farming the work out to Chinese factories.

"We want to bring jobs back to the United States," Walker told me when I visited him in Mississippi in January 2022. "But most factories say, 'We can't do it.' We've lost all that industry since the 1980s. No one is willing to do what we need done."

A factory in Georgia fielded Glo's requirements for a promotional box that could display the dolls, opening like a children's pop-up book.

"Their packaging engineers said, 'It's too complex,'" Walker recalled. "'You should just get this made in China.'"

In North America and Europe, the contemporary political discourse tends to describe China less like an enormous country with a rich and complex history than a cartoonish villain hell-bent on gobbling up livelihoods and resources. Its very name, *China*, gets tossed around by politicians and journalists as shorthand for seemingly every conceivable economic affliction, from mass unemployment in Rust Belt factory towns to the debt crises ravaging poor countries.

The discussion frequently indulges the same racist tropes that have animated Western treatment of China for centuries. In the 1870s, mobs of white, working-class Americans violently attacked Chinese laborers who had arrived to take jobs building the railroads, depicting them as mortal threats to wages, working conditions, and Christian values. Today, similar conceptions of the Yellow Peril are at play in portrayals of Chinese factory workers as robotic tools of a monolithic society intent on global domination.

Left out of such caricatures is a clear-eyed analysis of who has actually benefited from the interface of Western business and Chinese labor.

The consistent answer: the international investor class.

The Chinese workers who constructed the railroads in the nineteenth century did not represent an invasion. Rather, they were recruited by the industrialists who monopolized the profits of the age. The railroad magnates pressed Chinese workers into dangerous jobs at meager pay as a way to press down wages, and precisely so they could keep more of the profits for themselves and their financial backers—investors as far away as Europe.

In a similar vein, the factories that have exploded across China

over the last four decades have frequently been financed by investments delivered by multinational brands. Typically lost in the caus-
tic descriptions of a job-destroying Chinese juggernaut is the reality
that these international corporations and their shareholders have
pocketed many of the gains.

In claiming the White House in 2016, Donald Trump indulged
his trademark inflammatory rhetoric in advancing the notion that
China was synonymous with the forces downgrading American living standards.

"We can't continue to allow China to rape our country, and that's
what they're doing," Trump said at a rally in Indiana that spring. "It's
the greatest theft in the history of the world."

That indictment ignored the fact that American companies were
eagerly using Chinese factories to make their products, sending
their stock prices higher, while satisfying consumer demands for
low-priced goods—an antidote to the rising cost of living. Still, such
characterizations resonated, especially in white working-class communities, where a surge of Chinese imports had eliminated jobs.

Once in office, Trump installed as his chief trade advisor an obscure economist named Peter Navarro, whose primary credential
appeared to be having authored a hyperbolic book titled *Death by
China*. Together, they marshaled a trade war that featured across-
the-board tariffs on Chinese goods, depicting the nation's exports as
a threat to national security.

But the us-against-them rhetoric elided a crucial detail. It was
not China that was ransacking American sustenance. If a crime had
been committed, it was an inside job. The executives of American
corporations had used Chinese factories as a way to avoid paying
middle-class wages to workers in their own country while hoarding
the savings.

The enduring appeal of low-wage Chinese laborers to the international investor class was something that China's own leaders understood keenly as they set in motion the process that eventually
turned the country into the Factory Floor to the World.

Beginning in the 1980s, China's leaders began courting foreign investment that would advance the nation's industrial capacity. This would help them rescue the country from its mortifying state of backwardness and isolation, amassing wealth while leaving behind centuries of scarcity and vulnerability.

Modern Chinese history was the story of one catastrophe after another. An inherently great power that had produced papermaking, gunpowder, and the compass had been reduced to pauper status through centuries of colonial degradation, domestic turmoil, and extremist social experiments.

The British arrived in the nineteenth century, unleashing their gunboats to force a lucrative trade in opium. The Americans, the French, the Germans, and the Portuguese all carved out colonial entrepôts. The Japanese descended in the early part of the twentieth century, capturing vast swaths of Chinese territory while sowing terror.

The Communists took power in 1949, led by Mao Zedong, whose ardor for permanent revolution yielded a litany of calamities. The Great Leap Forward of the late 1950s—a preposterous reach for instant industrialization—ended in a famine that killed 30 million people. The Cultural Revolution that began the following decade killed as many as 8 million more, as Mao exhorted his disciples to attack his rivals.

But the late 1970s produced a fresh trajectory under the leadership of Deng Xiaoping, a pragmatist in a regime rife with battles over ideological purity. In a bid to spur innovation and growth, he undertook experiments that had once been anathema. No longer would the state, under Communist Party domination, set all prices for goods and commodities. A fledgling crop of entrepreneurs was permitted to make and sell products for profit.

Deng designated four so-called special economic zones in cities in southern China—three of them in Guangdong province, which sat next to Hong Kong. There, he allowed local officials to cultivate ventures with multinational companies and construct infrastructure.

Foreign investment poured into China, growing from about $5 billion a year in the late 1980s to beyond $30 billion by the middle of the following decade.

Honda began making motorcycles in China. General Motors and Volkswagen set up auto plants. Guangdong province alone saw its exports multiply from $2.9 billion per year in 1980 to $50 billion by 1994, its factories responsible for 40 percent of China's total exports. Over that same period, China's economy expanded by nearly 10 percent per year, an extraordinary pace.

In Shenzhen, one of the special economic zones, dirt roads became grand thoroughfares lined with luxury hotels and high-rise apartments adorned with aspirational names—Paradiso, City's Elite Homestead, Galaxy International Park. Migrant workers carrying meager belongings in pails balanced from mop handles tromped past spa resorts stocked with orchids and Jacuzzis, on their way to soot-stained factories where they slept eight to a dorm room, their laundry flapping from the windows.

The factories of Shenzhen were soon making three of every four artificial Christmas trees sold in the United States. A single complex in the center of Guangdong manufactured 15 million microwave ovens per year, or about 40 percent of those sold on earth.

China's export-led economic transformation was perhaps the greatest antipoverty program in human history. From the beginning of Deng's economic reforms in the late 1970s until the new millennium, the number of Chinese who were officially poor plunged from nearly 900 million to fewer than 500 million.

And that was before the landmark event that would dramatically accelerate the country's emergence as a global economic power—the 2001 deal that brought China into the ranks of the World Trade Organization. More than anything else, that agreement helped explain why Hagan Walker and countless other American business chiefs found themselves consumed with events playing out on the other side of the world.

China's inclusion in the global trading system was championed by multinational companies as a way to dramatically reduce their costs. But the executives of the largest companies and their political allies realized their goal through a savvy laundering of their aims. In their telling, adding China to the WTO was about promoting democracy.

No one proclaimed this message more emphatically than Bill Clinton, the president who ultimately brought China into the WTO. And no one better personified the pungent mix of shameless opportunism and idealistic rhetoric that characterized American dealings with the emerging superpower across the Pacific.

Even for a figure inclined toward impromptu displays of showmanship, Bill Clinton outdid himself inside the massive banquet hall in the heart of Beijing.

It was a balmy night in June 1998, and the president was in the midst of alternately wooing and pressuring China's government to assent to American terms on a deal bringing the nation into the World Trade Organization. He and then–First Lady Hillary Rodham Clinton were attending a state dinner at the Great Hall of the People, the colonnaded fortress occupying the western edge of Tiananmen Square.

Only nine years earlier, the square had been the locus of an extraordinary protest movement led by students who demanded greater freedom and a crackdown on corruption. In a spectacle broadcast worldwide, some of the demonstrators had erected a replica of the Statue of Liberty—a courageous display of democratic yearning.

The People's Liberation Army ultimately put down the uprising with a massacre that killed several hundred people. That act of brutality had defined China in international discourse for years after, rendering the country off-limits to many North American and European businesses. But China's pariah status was already fading as Western executives salivated over the lucrative opportunities

waiting to be exploited there. The Clintons were in Beijing in the service of that cause.

Clinton had initially presented himself as one of China's most
strident critics. During his 1992 campaign, he attacked the incum-
bent president, George H. W. Bush—a former American ambassador to China—for normalizing trade relations. He accused Bush of
pandering to the "butchers of Beijing." As Clinton described it, Bush
was guilty of having "let his friendships in China obscure what those
kids did in Tiananmen Square."

Things would be different in his own administration, Clinton
promised. "Observe human rights in the future," he warned Chinese
leaders, "open your society, recognize the legitimacy of those kids
that were carrying the Statue of Liberty." Otherwise, prepare to be
cut off from the American marketplace.

Chinese leaders chafed at being lectured to on human rights by
sanctimonious politicians from the United States, a former colonial power that had amassed wealth through legal slavery. In more
recent times, the supposed holiness of human rights had not dis-
suaded the American military from carpet-bombing Cambodia or
dropping napalm on Vietnamese children, displaying its moral code
in what was essentially China's backyard.

Clinton had himself opposed the Vietnam War, but he did not let
the unsavory aspects of his own country's history brunt the force of
his moral denunciation. Soon after taking office, he signed an executive order threatening China with tariffs as high as 70 percent
unless it showed "overall significant progress" on human rights.

So much for all that.

On this night in June 1998, inside the citadel of Chinese power,
Clinton demonstrated the updated values guiding his administration's approach to China.

The demonstrators had been gunned down only a few hundred
yards from the banquet hall, but that event had seemingly been dispatched to ancient history. The Clintons posed for pictures. They sa-

luted their hosts, Chinese president Jiang Zemin and his wife, Wang Yeping.

"The American people admire the great strides China has taken," President Clinton declared during a predinner toast. "We Americans appreciate the mutual respect of our relationship."

After a meal featuring shark's fin and grilled beefsteak, the two couples worked their way to the far end of the enormous banquet hall. There, they greeted the musicians who had been entertaining them throughout their dinner—members of the People's Liberation Army Band.

To the delight of the assembled dignitaries, President Clinton wielded a baton and briefly conducted the band as it played a John Philip Sousa march, "Hands Across the Sea."

Clinton's turn as maestro before the very institution that had slaughtered peaceful protesters marked the completion of a striking makeover. He was now leading a campaign orchestrated by American business interests to ditch commerce-impeding considerations of human rights in favor of exploiting China as the holy grail of profit centers. Getting China into the WTO was the centerpiece of that effort.

Headquartered in Switzerland, on the shores of Lake Geneva, the WTO was the referee for international trading disputes. Its 160-plus member nations were supposed to abide by a common set of rules, submitting to an adjudicatory process when squabbles arose. Members pledged to treat one another equally, opening their markets to goods free of barriers, while extending the same terms of trade to all.

China wanted in for the simple reason that this would unlock fresh export opportunities and lure more foreign investment. But gaining entry required that China secure the blessing of two-thirds of the WTO's existing members.

China had to open its markets to foreign competition, a prospect that horrified the heads of China's industrial giants. But Premier

Zhu Rongji, then spearheading economic policy, saw the mere process of applying to join the WTO as an accelerant in remaking the Chinese economy.

Into the mid-1990s, state-owned companies still employed roughly 70 percent of urban Chinese workers. More than 80 percent of the outstanding loans in the Chinese banking system were going to such firms, many of them insolvent. These zombie companies were monopolizing finance that might have otherwise nurtured nimbler, more innovative entrants—the sorts of factories that would eventually attract giant retailers like Walmart and niche players like Glo.

The greatest resistance to change came from local Party officials. They feared the anger of the populace when state-owned employers went bust. They frequently enjoyed a cut of the bank loans that kept such companies in business—an incentive to keep the money flowing.

Under the terms of its WTO entry, China agreed to pull back subsidies for state-owned companies. China's international obligations would take precedence over the intransigence of small-minded Party leaders.

Zhu was aided in his mission by a crucial constituency—the American business lobby, which recognized a gold mine in China. Here was a country of 1.2 billion people and counting, making it the largest potential market for practically everything.

Telecommunication companies and financial giants stood to gain clearance to buy larger stakes of Chinese ventures. Hollywood aimed to export its films. Software companies would get a crack at the Chinese market. Agribusiness would be free to sell mountains of soybeans, wheat, and other crops.

Multinational brands were especially enthralled by the savings to be gained as they moved manufacturing away from wealthy nations with unions and workplace safety regulations, transferring production to plants in China. There, Party officials were the law, and their loyalties could be purchased with a cut of the bounty.

Unions were banned, and hundreds of millions of rural people were streaming toward cities in desperate pursuit of work—an ideal recipe for low-cost production.

Here was an ironic confluence of interests. The People's Republic of China—still dominated by a Communist Party whose power flowed from a peasant-led revolution—beckoned as the ultimate joint venture partner for Walmart, Apple, and other icons of American capitalism.

Still, multinational companies were fearful of seeing their designs on China construed as a crude reach for profit. Stories of pervasive Chinese repression, human rights abuses, and labor exploitation posed reputational risks. So, American business associations and their political allies reframed the terms of engagement, while settling on a consensus narrative: expanded trade with China would advance freedom.

Yes, American companies stood to make enormous amounts of money in China, but that was incidental to the transformational mission at hand. A chorus of Clinton administration officials declared this repeatedly. Once China was inside the WTO, its prosperity would depend on operating within its norms. Its reforms would pass the point of no return, changing China forever.

Chinese companies would require greater understanding of the outside world to capitalize on the benefits of joining the global trading system. Chinese consumers would gain exposure to international fashion trends, music, entertainment, and sports, along with the latest technological gadgets. All of this would infuse Chinese society while generating demand for the ultimate Western export: democracy. First, Chinese people would get a taste of Kentucky Fried Chicken. Then, they would demand the ballot box.

"By joining the WTO, China is not simply agreeing to import more of our products; it is agreeing to import one of democracy's most cherished values: economic freedom," Clinton declared in a speech on the eve of a key congressional vote that cleared the deal. "The genie of freedom will not go back into the bottle."

He added a dose of Silicon Valley techno-hype: "In the new century, liberty will spread by cell phone and cable modem."

There was logic to this construction, even as there was ample cause to doubt the Hollywood ending. Markets demanded intelligence, and repressive governance was at odds with the free flow of information. Expanded trade would presumably force China's leaders to choose between protecting their chokehold on the political sphere or advancing economic development.

But the most powerful driver of such faith was the imperative of the bottom line. The greatest beneficiaries of China's entry into the international trading system were the shareholders of the multinational brands that would exploit China as a source of cheap workers.

In Washington, lobbyists representing the largest American companies described China's entry to the global trading system as an opportunity to advance the agenda of the reformists in Beijing. Clinton's Treasury secretary, Robert Rubin—who would go on to head the financial services giant Citigroup as it penetrated China— assured Congress that the country's entry to the global trading system would "sow the seeds of freedom."

Trade had been a foundational piece of American foreign policy since the end of World War II. The Clinton administration put stock in the notion that trade could be the linchpin of peaceful relations with China. Here was a means of limiting the risk of hostilities over Taiwan, the self-governing island claimed by China and defended by the United States. The more that China's fortunes became dependent on international commerce, the more it would have to lose from the economic shock that would surely follow an attack on Taiwan.

Yet behind the scenes, the leaders of American corporations were advancing self-interested arguments. Major lobbying organizations from the National Association of Manufacturers to the Business Roundtable asserted that tying trade policy to human rights rendered their dealings in China uncertain, limiting their appetite to invest. And that was impeding their grandest plans—selling products to Chinese consumers while relying on Chinese factories as central components in their supply chains.

As Clinton's secretary of state, Warren Christopher, later put it: "The business community had convinced the president that trade for America was a higher value, or perhaps, to put it more charitably, that nothing would be accomplished in the field of human rights by denial of trade, and so that became the basic policy."

This chain of events, to a great extent, explained how manufacturing shifted from factory towns in places like Michigan and Indiana to the industrial enclaves of coastal China. And this was why, two decades later, Hagan Walker found himself seeking factories in China to make his products. Because his home country no longer had the capacity to satisfy his needs.

By the time Walker placed his order for the *Sesame Street* figurines, the idea that China's entry to the global trading system would advance freedom had come to seem like cruel farce.

The permanent muting of the debate over human rights indeed opened the floodgates to a new wave of foreign investment, which immediately surged beyond $50 billion a year. Beijing ensured that Chinese industry possessed the infrastructure needed to grow into an enormous export power. Some $14 billion in investment was directed at seventy-two privately operated ports between 1990 and 2013, along with billions more into government-run facilities. China's government oversaw the expansion of the Chinese maritime industry, lavishing state bank loans plus direct subsidies worth $132 billion on ocean carriers, shipbuilders, and other industry concerns between 2010 and 2018.

But the government harnessed China's economic strength to solidify its domination of Chinese society while ratcheting up territorial confrontations with its neighbors. Much of the nation's wealth was directed toward building out its military capability. China's leaders took the technological developments that Clinton celebrated as purveyors of liberty and deployed them as the building blocks for an Orwellian surveillance apparatus. The state directed those capabilities toward the systematic repression of ethnic minority Uyghurs in the province of Xinjiang. It extended its mode of control to social media and intensified its harsh punishment of dissidents. It imposed a draconian security law on Hong Kong, breaching an agreement to respect the city's freewheeling ways under a treaty with Britain that returned the former colonial trading post to Chinese control.

But one part of the promised transformation came to pass. China became an integral part of the global supply chain. The value of China's exports multiplied from $272 billion in 2001—the year it officially joined the WTO—to more than $3.5 trillion twenty years later.

Chinese companies were making roughly 80 percent of the world's air conditioners, 70 percent of all mobile phones, and more than half of all shoes. And Chinese industry was increasingly engaged in far more advanced pursuits like aviation, biotechnology, and computer chips, while manufacturing 80 percent of the world's solar panels.

In North America and Europe, much of this development was construed as nefarious. Chinese industry was manipulating the rules of the global trading system. It was tapping state-owned banks and emerging stock markets to steer capital to politically connected companies. Chinese officials looked the other way on ventures that dumped poisonous chemicals into rivers—another advantage over companies that had to answer to strict environmental rules in their home countries. Because the Chinese press was heavily censored, wrongdoing was easily hidden from public view. And when foreign companies invested in China, they frequently suffered the theft of their intellectual property, watching counterfeit versions of their products pop up in markets throughout the country and around the world. Not even pharmaceutical companies were immune to this reality. In short, Chinese industry was not playing fair, while capturing jobs from hapless competitors that had to answer to unions, regulations, journalists, and basic decency.

There was validity to these laments. Chinese companies were indeed undercutting labor and environmental standards to seize market share. And this was coming at the expense of jobs in other countries. A decade after China's accession to the WTO, nearly 1 million American manufacturing jobs had been eliminated directly by Chinese imports. And that number roughly doubled after factoring in the impacts on communities that saw factories disappear—the restaurants that lost customers, the truck drivers no longer needed, the plumbers and electricians whose incomes fell as people moved away. Politicians from both major parties came to describe China as a malevolent force.

One could be tempted to describe China's gains as a transfer of wealth from places like the American Rust Belt to industrial areas in Guangdong province. But such framing skipped over the fact that the primary beneficiaries of China's export boom sat in the United States itself—not in fading factory towns, but in boardrooms in New York and Seattle.

The microwave ovens, the Christmas ornaments, the televisions—all were being sold on the shelves of the largest retailers, dropping their costs while boosting their sales. Their executives were prospering mightily as their stock prices rose.

American businesses justified the bargain by taking credit for supplying consumers with cheap products. During a hearing in Washington to debate whether the United States should normalize trade relations with Beijing—a step that cleared the way for WTO entry—members of Congress spoke darkly of forced sterilization and slave labor in China. They also heard from Clark A. Johnson, chief executive officer of Pier 1 Imports, a purveyor of rugs, furniture, and other home décor. He was speaking on behalf of the National Retail Federation, a powerful trade association that represented 1.4 million businesses.

"The mission of a retailer in America is to provide an assortment of a wide range of goods from everywhere in the world," Johnson said. "China makes products that working families can afford."

Those words carried the day. And they revealed a central truth behind the ceaseless transferring of industrial capacity from wealthier countries to China: because consumers essentially demanded it.

You could survey the populace on sentiments about human rights, fair pay, and environmental imperatives. You could debate the moral and economic virtues of dismantling factories in North Carolina while transferring production to China. Such conversations involved nuance and trade-offs. But the consumer wandering the aisles at Target or perusing the offerings at Amazon rarely paused to grapple with such concerns. The lowest price frequently trumped all other considerations. And making products in China was generally the best way to deliver lower prices.

The growth of Chinese imports boosted spending power in the typical American household by 2 percent, or about $1,500 per year, between 2000 and 2007, according to one study. By this calculation, every factory job surrendered to trade with China in those years was compensated by an increase of $400,000 in consumer spending capacity. Chinese-made products pressed down the prices on a range of American goods by 0.19 percent per year between 2004 and 2015, another study found.

Walmart, a company ruled by a near-religious zeal for low prices, took China's entry to the WTO as the impetus to establish a global procurement center in Shenzhen, the place where it coordinated its buying of products worldwide.

When I visited the center in early 2004 for a story on the company's expansion, representatives from dozens of factories across China—each hoping to secure a contract to manufacture one of Walmart's products—were assembled in a giant waiting room. They sat in uncomfortable wooden chairs, sipping tepid tea out of flimsy plastic cups, as they waited for hours to meet a Walmart buyer. Walmart had positioned itself to demand whatever price it desired, aided by the implicit threat that should one factory balk, another could immediately be summoned from inside the same waiting room.

Frequently, local factories agreed to make a product, and then prepared to deliver—taking on the expenses of hiring workers and buying materials—only to confront fresh demands from Walmart for even lower prices, often below their costs. Typically, managers assented while recovering some of their losses by stiffing workers.

Walmart's global database of suppliers then included six thousand factories. More than 80 percent were in China. By 2003, only two years after China entered the WTO, Walmart was spending $15 billion on Chinese-made products, a sum that encompassed almost one-eighth of all of China's exports to the United States.

American consumers were getting rock-bottom prices at Walmart superstores. American factory laborers were losing their jobs. Chinese factory workers were getting cheated by bosses who themselves were getting squeezed. Meanwhile, the descendants of Walmart's founder, Sam Walton, saw their collective fortune swell beyond $200 billion, making them the wealthiest family on earth.

Part of the reason that American public opinion and the political class eventually swung so sharply against engagement with China stemmed from bitterness over the bogus promises that had been made about democratization as a reason for expanded trade. But part of the anger reflected an American failure to cushion those harmed by Chinese imports.

A federal program called Trade Adjustment Assistance was supposed to provide factory laborers made jobless by low-priced imports with cash assistance and training for other work. But Congress routinely refused to adequately fund the program, even as it managed to finance tax cuts for billionaires and corporate interests. As a result, fewer than one-third of those eligible for benefits in 2019 actually received support.

But these sorts of details rarely got mentioned in a political conversation long on accusations of perfidy and short on substance. In a triumph of crude political messaging over the complex accounting of trade, the electorate broadly came to believe that Chinese industry was solely a predatory force. "There has been an overcorrection in the narrative, which is that we were just taken advantage of," Jessica

Chen Weiss, a China expert at Cornell University and a former State Department official in the Biden administration, told me. "We didn't do a good job of distributing the benefits, but they were nonetheless real."

Within manufacturing, the momentum toward China was self-reinforcing. The more that companies shifted production to China, the more the prices they secured there were insinuated into the workings of international commerce. That shaped collective understandings about the value of everything—an electronic component, a coil of steel, an hour of labor.

In a marketplace in which many companies were already making goods in China, the company that failed to do likewise risked distinguishing itself as uncompetitive.

This was why industrial production exploded in China just as American factories disappeared.

And this was why Hagan Walker could not locate a plant in the United States that could bring his creations to life at anywhere close to the prices prevailing across the Pacific.

So he did what everyone else did. He sent his business to China.

CHAPTER 2

"EVERYONE IS COMPETING FOR A SUPPLY LOCATED IN A SINGLE COUNTRY."

The Pandemic Reveals the Folly

Over the years, Glo had worked with multiple factories in southern China to make its cubes, with mixed results. The prices were consistently low. But the quality varied.

Once, Walker had received a shipment of one hundred thousand cubes from a supplier in Guangdong, only to discover that nearly one-third were defective. He and Barker sat in a conference room for days, testing the cubes by dropping them into water one by one. They had been forced to disappoint customers, delaying shipments while waiting for another batch to come in.

On the *Sesame Street* order, such an outcome would be calamitous. They needed a factory operation that could deliver.

In late 2020, Walker reached out to an operation in the city of Ningbo, in Zhejiang province. It was part of a collection of factories that operated under the brand Platform88. The company was overseen by an American, Jacob Rothman, who had spent most of his adult life in China.

Raised in Southern California, Rothman had studied religion at Bowdoin College in Maine, contemplating a life as a rabbi. Instead, he had joined the family business, a broom and mop factory in Stockton, California, sixty miles east of San Francisco.

He had first gone to China in 2000 to locate factories that could manufacture the company's wares cheaply. He had stayed on as the business pivoted to more lucrative pursuits, supplying products for men's gift catalogues—fishing gear, barbecue accessories, hot shaving cream dispensers.

Eager to master Mandarin and the social mores of China, Rothman eschewed the expatriate scene. He stayed in small, local hotels rather than the five-star palaces filling every major city. Soon, he launched his own company, joining forces with a Chinese partner.

His six factories churned out products for major multinational retailers, including Walmart and Weber, the maker of backyard barbecues. Rothman's engineers helped tweak the specifications for the best results, and his procurement people could find needed materials from across China, from specialty plastics to LED lights.

Walker liked the sound of this. He began talking to Rothman's team about proceeding with the *Sesame Street* order.

By then, of course, Walker, Rothman, and the rest of the international business world were contending with an enormous new variable.

A novel coronavirus known as COVID-19 was working its way around the globe, spreading death, fear, and dysfunction.

The pandemic had begun in China. The first cases had emerged a year earlier, in December 2019, in the industrial city of Wuhan, home to more than 11 million people. Local authorities initially covered up reports of a mysterious ailment that was causing pneumonia-like symptoms.

A month later, with the coronavirus raging to the point that it could not be suppressed, the Chinese government imposed the first of a series of comprehensive lockdowns, isolating Wuhan and surrounding Hubei province from the rest of humanity by canceling all flights and trains departing the area.

By February 2020, the authorities were forcibly rounding up people suspected of carrying the coronavirus and confining them to makeshift quarantine facilities.

As some 300 million migrant workers across China returned to rural homes to celebrate the Lunar New Year, many found themselves stranded. A confounding patchwork of restrictions imposed by officials from place to place—barriers blocking highways here, shutdowns of train stations there—made it impossible for many to return to jobs.

Thousands of factories were ordered shut to prevent the spread of the coronavirus. Even those that were allowed to open contended with crippling shortages of workers and parts. Over the first two months of 2020, China's exports plunged by 17 percent.

For the rest of the world, the mass disruption of Chinese industry posed enormous consequences. China was the dominant source for a vast range of parts and materials that were critical to the making of products on nearly every shore.

Modern manufacturing had come to rely on an extraordinarily complex global supply chain comprised of parts and raw materials drawn from scores of countries. Cargo ships were no longer just bearers of finished products. Increasingly, they were conduits for components that could be processed in one place—coated with a chemical treatment here, pounded into a desired shape there—and then combined with other parts somewhere else, to be assembled into finished goods.

The typical car required an average of thirty thousand individual parts forged in plants around the globe, with many components transferred multiple times across oceans to be combined with others before the final assembly of the vehicle. If just one of these parts could not be purchased, the rest were effectively stuck in limbo, with final production impossible. Apple purchased parts from suppliers in forty-three different countries across six continents, while fitting them together into finished products at factories in China and Taiwan. Even seemingly simple products like disposable diapers required more than fifty materials.

Hubei province, the epicenter of the pandemic, was an especially important link in the supply chain. Some fifty-one thousand companies worldwide had a direct supplier somewhere in the territory. At least 5 million businesses depended on suppliers that themselves sourced parts or raw materials from Hubei.

Suddenly, the world absorbed the pitfalls of having entrusted so much of its production to a single country.

By February 2020, Apple was disclosing that it would miss targets for the shipment of its latest iPhones as its Chinese factories contended with disruption. Toyota grappled with a dearth of parts as workers at its plants in China were stuck in quarantine. Fiat Chrysler, another automaker, prepared to shut factories in Europe because of an inability to secure parts made in China. The French automaker Renault halted production in South Korea, citing a shortage of parts brought from China. So did Hyundai, the world's fifth largest automaker.

Fashion companies scrambled to shift orders from China to Turkey. Toy retailers, who collectively relied on Chinese factories for 85 percent of global production, girded for delays and shortages.

Nintendo delayed deliveries of its popular gaming console, the Switch. The gadget was made in a factory in Vietnam, but production depended on parts made in China. From India to Japan, manufacturers depended on China for nearly two-thirds of their imported electronics components.

A lingerie company that had moved its operations from China to Bangladesh to sidestep Trump's tariffs on Chinese exports nonetheless suffered a hit to its production. It was stuck waiting for straps and bra cups still made in Chinese factories.

The toilet paper shortage that would bring the supply chain disruption home to ordinary people around the world was mostly a story of panic buying and hoarding. Images of empty shelves shared via Facebook prompted people to run out and lock up a stash. But part of the trouble stemmed from the supply chain. Ten percent of the giant rolls of paper used to make bathroom-size rolls were produced in China and India.

Hospitals contending with the first waves of COVID-19 from Europe to North America found that they could not purchase enough protective gear for their doctors, nurses, and other medical workers. Much of the global supply was manufactured in China—80 percent of face masks alone—and specifically in Hubei.

In New York, hospitals and nursing homes were relying on a rapidly diminishing government stockpile to outfit their people with protective gear. Italian authorities were telling hospitals to prepare to reuse face masks given their inability to secure more stocks from China. In Sweden, thousands of elderly people died in nursing homes—many placed on morphine drips without so much as an examination by a doctor—in part because of severe shortages of face masks and gowns for staff.

Ventilators became so scarce that American auto plants were enlisted to reconfigure their operations to make them. Part of this desperation reflected how, before the pandemic, the United States had imported 17 percent of its ventilators from China.

Pharmaceutical manufacturers could offer no reassurances that their own stocks would withstand the crisis. Chinese factories were the source of 90 percent of many antibiotics sold in the United States, and 70 percent of basic medicines like acetaminophen. Chinese manufacturers dominated the supply of important ingredients used by pharmaceutical manufacturers in India—itself a major producer. And Chinese plants were the largest suppliers of basic chemicals used to make a litany of generic medicines used to treat patients hospitalized with COVID-19.

Fear combined with rising nationalism to yield a frenzied scramble for the raw materials needed to make drugs and, eventually, vaccines. Dozens of countries enacted export bans.

At the center of the conflict was the reality that many of the needed goods were made in China.

By early April 2020, 70 percent of American hospitals and pharmacies were reporting shortages of at least one drug used to treat coronavirus patients.

"Everyone is competing for a supply located in a single country," said Rosemary Gibson, a health care expert at the Hastings Center, an independent research institution in New York.

Hagan Walker did not need advanced computer chips or ingredients for pharmaceuticals. He was not making ventilators or medical-grade face masks. He was simply manufacturing plastic figurines that lit up with the aid of basic electronics.

And still, he was at the mercy of the turmoil upending Chinese industry.

His supply of basic materials depended on the operations of scores of factories that produced the components for his products—the chemicals used to make plastic, the basic metals for electronics, the cardboard for his packaging. All of these were now scarce. So long as elements of Chinese industry were disrupted, his supply was threatened, impeding his ability to produce his wares.

Here was a worry that went back to the beginning of mass production more than a century earlier. The more a factory was dependent on outside players to supply its parts and raw materials, the greater its vulnerability to trouble.

Henry Ford, the godfather of the modern assembly line, had seen all this coming. His struggles to avoid the pitfalls of elaborate supply chains reveal the roots of the crisis that has assailed the globe in more recent times.

———

Self-assured and ruled by grand ambitions, Ford was not one to ruminate on inner doubts. Yet by the summer of 1915, the complexity of his commercial empire was causing him agita.

His company had already claimed distinction as a triumphant pioneer in the fledgling automobile industry. His Model T—the first car produced at scale on a moving assembly line—was in the process of transforming the American way of life.

The car was reconfiguring geography, opening the spaces beyond cities for exploration, and creating fresh opportunities for housing, work, and leisure. It would enable decades of suburban development and spur the construction of a continent-wide network of highways.

In a nation that celebrated mobility with near-mystical reverence, Ford had succeeded in pushing out the contours of freedom itself.

And as he expanded his sights globally—eventually erecting factories in England, Egypt, South Africa, and Japan—Ford was on his way to becoming perhaps the most famous person on earth, his words mined for prosperity-enhancing wisdom much like those of Steve Jobs and Jeff Bezos decades later.

Yet he was tormented by worries about variables that were beyond his control. One question gnawed at him without relief: how could he ensure access to enough supplies to produce all the cars that he envisioned?

Ford was dependent upon outsiders for the raw materials and parts required to make the vehicles that bore his name. His demand for these wares was growing in tandem with his production, testing the outer limits of what his suppliers could deliver.

The Model T, introduced in 1908, was the product of the commercial ecosystem of Detroit. Perpetually short of capital, Ford had initially relied on local companies to make his chassis, wheels, tires, and engines. He depended on railroads and ships that plied a series of rivers and canals to summon the coal that powered his furnaces, leaning on the same vessels to transport finished vehicles to a growing network of dealers across the country and, eventually, around the globe.

He had fashioned his assembly line inside his Highland Park factory, a once-cutting-edge facility that had begun production in 1910. Five years later, Ford had become convinced that the plant had reached the heights of its potential productivity.

He was also dismayed by slowdowns, as suppliers withheld key materials in negotiations over sales terms.

As Ford remarked to a business associate, he was intent on avoiding being "dependent to the point where he could be pinched."

So, he began quietly amassing two thousand acres of Michigan marshland alongside a silty thread of water known as the River Rouge while preparing to develop what would become the world's largest industrial complex.

"We aim to make some of every part so that we cannot be caught in any market emergency or be crippled by some outside manufacturer being unable to fill his orders," Ford would write in his memoir.

The plant, known as the Rouge, was a monument to the emerging concept of vertical integration. All the processes and stages of turning raw materials into a finished car would be maintained within its confines.

Ford chose the site for its proximity to an expansive network of competing railroads, diminishing his reliance on any single operator whose managerial shortcomings, greed, or monopolistic tendencies might stymie his operations. He dredged the river to create a berth with room for large ships to dock, relying on the water to expand his menu of suppliers.

"His blueprint called for a vast extension of the manufacturing process in order to construct an entire vehicle, from raw materials to finished product," wrote Steven Watts in his seminal Ford biography, *The People's Tycoon*. "The complex would include not only an assembly plant, but a port and a shipyard, a huge series of storage bins, a steelmaking operation, a foundry, a body-making plant, a sawmill, a rubber-processing facility, a cement plant, and a power plant. Thus, the goal of the Rouge was not just a matter of sheer size but complete control of production, from the time iron ore came out of the ground until a Model T rolled off the assembly line."

Ford's pursuit of self-sufficiency would lead him to extremes that sometimes ended disastrously. His foray into the rainforest of the Amazon basin, where he erected a town that was supposed to produce rubber, proved an embarrassing debacle.

But his obsession with safeguarding the resilience of his supply chain would remain a hallmark of his oversight of his company—as well as an approach that has been vindicated by our times.

He correctly divined that large-scale manufacturing would force producers to scrutinize their supply chains and protect themselves from vulnerability to transportation breakdowns and other unforeseen problems, from natural disasters to the greed of middlemen.

He also recognized that the investors who backed his ventures frequently did not share his concern for provisioning against trouble. Wall Street money managers were hungry for immediate gratification—for dividends that came at the expense of the capital that Ford preferred to invest in expanding and upgrading his factories. Battles for control with his financial backers proved an unending feature of Ford's tenure.

Despite his popular acclaim in his own era, Henry Ford is not fit for lionizing today. He extolled the traditional, church-on-Sunday values of his rural Michigan roots in terms infused with white supremacy. "The trouble with us today is that we have been unfaithful to the White Man's traditions and privileges," he wrote in an essay. His disgust for financiers went hand in glove with a virulent strain of anti-Semitism. He fired workers who tried to unionize his plants and unleashed brutal violence to suppress their mobilization.

And still, a reexamination of Ford's business legacy yields insights about the perils of organizing society around shareholder interests to the exclusion of common sense—lessons that remain relevant today.

Born in 1863, Ford was supposed to be a farmer. His father, a refugee from the potato famine in his native Ireland, had established a home on the fertile soils of Michigan. Yet at thirteen, while riding into Detroit on a horse-drawn carriage, Ford encountered a spectacle that changed his life—a portable steam engine that was hauling a threshing machine. Transfixed, he jumped out of the carriage and interrogated the driver on the particulars of his contraption.

Three years later, he abandoned the family farm and walked nine miles to Detroit to seize his future.

Detroit was already a leading hub of American manufacturing, owing to its centrality in the Great Lakes region—a natural transportation corridor. Local factories were making railcars, ovens, and stoves, among other modern conveniences. Ford hired on as an apprentice at a factory that shaped parts out of brass and iron. By his late twenties, he was consumed with developing a horseless carriage.

He rolled his first model through the city streets on a thrilling maiden voyage in 1896. Later that year, at a professional dinner held on New York's Coney Island, he secured a seat near the guest of honor—the legendary inventor Thomas Alva Edison.

Edison's creations had already altered the shape of the modern world. He had produced a light bulb, the phonograph, and a movie camera. Ford was starstruck. He regaled Edison with accounts of his experiments with gasoline-powered engines.

The next day, Edison invited Ford to accompany him to Manhattan by train. As they clattered through Brooklyn, the two men discovered a natural affinity that would sustain a lifelong friendship. What they discussed most fervently was their ceaseless worries about finding components needed to turn their visions into reality.

"We talked mostly of the difficulties of obtaining the right kind of materials and supplies in the working out of new inventions," Ford related. "I told him that for my first car I could find no suitable tires and had to use bicycle tires, and he told me something of the trouble which he had met in finding suitable bulbs for the incandescent light and how he had to have them blown himself."

Here were two of the most celebrated innovators in American history, forging a bond through the shared acknowledgment of an inescapable truth. Without a reliable supply chain, their inventions were just dreamy notions.

That insight would guide Ford for the rest of his days, as he pursued his ultimate ambition—producing an automobile for the mass market.

"I will build a motor car for the great multitude," he declared, as he announced the creation of the Model T. "It will be large enough for the family, yet small enough for the individual to run and care for. It will be constructed of the best material, by the best men to be hired, after the simplest designs that modern engineering can devise. But it will be so low in price that no man making a good salary will be unable to own one."

The key to achieving that goal was to make as many cars as possible, spreading out the costs of production. And the means of bringing that about was to purchase materials in huge quantity.

He instituted a stock control department to oversee his inventory of parts.

Initially, making one Model T consumed a dozen hours. Eager to speed up the process, Ford examined other industries that had successfully automated. Slaughterhouses were carving cattle into steaks with the aid of moving rails hung from ceilings. Textile factories were deploying conveyor systems to accelerate their operations. Ford experimented with attaching ropes to chassis and pulling them across the floor, bringing the work to his laborers rather than having them move to the work.

But that was merely a glimpse of where he was headed. His

Highland Park factory, opened in January 1910, would become the showcase for the first assembly line used to produce automobiles.

Situated on the grounds of an old racetrack, the building featured enormous windows and a glass roof. In 1913, the first year his assembly line was in operation, Ford's production of Model Ts more than doubled, from 82,000 a year to 189,000. It was making more than half of all cars in the United States. By the summer of the following year, the plant was producing a finished Model T in a mere ninety-three minutes.

Scale worked as planned, allowing Ford to drop the price of a Model T from $950 in 1910 to $360 by 1916. Managers from factories around the globe descended on Highland Park to study this miracle of productivity.

With production rigorously routinized, Ford was able to precisely plan the flow of needed parts and materials. He achieved an early version of what the business world would later embrace as "Just in Time manufacturing," delivering parts to the assembly line as they were required.

"We have found in buying materials that it is not worthwhile to buy for other than immediate needs," Ford wrote. "If transportation were perfect and an even flow of materials could be assured, it would not be necessary to carry any stock whatsoever."

But transportation was far from perfect. Here was the spur for Ford's eventually buying his own railroad, harnessing it to carry in coal from his own mines in Kentucky and Virginia. He amassed his own fleet of ships, deploying them to deliver iron ore mined in Minnesota and Michigan's Upper Peninsula. He brought in timber from his own forests. He erected his own steel mill.

Ford was wrestling with the creation of what we now know as the supply chain. He had been especially alarmed by how the price of glass had climbed to "outrageously high" levels during World War I.

"We are among the largest users of glass in the country," Ford wrote. "Now we are putting up our own glass factory."

———

In September 1916, as Ford was readying plans for the Rouge, he found himself crosswise with some of his first investors, the Dodge brothers—themselves leading innovators in the emerging auto industry.

It had not escaped their attention that Ford's profits over the previous year had reached $16 million. The company had $50 million in cash stashed in the bank.

Ford was adamant that this money be used for expansion. The Dodge brothers were equally insistent that a generous helping of funds be served up to investors in the form of dividends.

Unable to gain an audience with Ford, the Dodge brothers resorted to a blunt form of communication. They filed a lawsuit. They petitioned the court for an injunction that would freeze Ford's expansion plans at the Rouge while forcing him to hand out three-fourths of his cash as dividends. The court assented.

Ford was taken aback. The Dodge brothers were imperiling not only his designs for the Rouge but the central organizing principle of his company.

"I do not believe that we should make such an awful profit on our cars," he declared on the witness stand during the ensuing trial. "A reasonable profit is right, but not too much. So it has been my policy to force the price of the car down as fast as production would permit, and give the benefits to users and laborers."

Under withering cross-examination, Ford declared that the very point of his business was to provide jobs and build affordable cars.

"Business is a service, not a bonanza," he said.

He lost the case. The court ordered him to distribute the dividends, though—via an appeal—he secured the right to go ahead with building out the Rouge.

Through the lens of history, the Dodge brothers' challenge stands out as a portent of what lay ahead.

Business would be run primarily for the enrichment of the shareholder, and to the exclusion of other interests—a reality that would eventually warp the supply chain.

On a drizzly morning in May 2022, I walked the perimeter of the Rouge, still in operation more than a century later. I looked down from a catwalk at the spectacle of hundreds of workers laboring on the assembly line, wielding tools as they fit together the pieces of Ford's most popular contemporary vehicle, the F-150 pickup truck.

Henry Ford was long gone. He died in 1947, a few years after suffering a debilitating stroke. Detroit had devolved into a poignant example of what happened when middle-class factory jobs were replaced by poverty-level positions at retailers, warehouses, and fast-food chains. Next to the old Highland Park factory—now a forlorn brick shell—a shopping center called Model T Plaza was anchored by a dollar store, a payday lender, and a plasma collection center awaiting people willing to swap blood for cash. The car itself had become an instrument of white flight.

But the Rouge plant was still humming, the logical extension of everything that Ford had set in motion.

The assembly line wound through the plant, bearing the chassis of the vehicles, as forklift drivers dropped required parts at the places they were needed, in accordance with a meticulously calibrated schedule.

Women now worked the line alongside men. Robotic arms bobbed back and forth, plucking black-tinted windows from a tray and affixing them to the backs of the truck cabs. Near the end of the line, robots trained lasers on the seams around doors and lights, probing to ensure that they were impervious to water.

Ford's company was under the nominal leadership of his great grandson, William Clay Ford Jr., who held the title of executive chair. But the real boss was Wall Street. The company's nearly 4 billion publicly traded shares were collectively worth more than $31 billion. More than half were controlled by enormous financial institutions like Vanguard, the leading purveyor of mutual funds, and BlackRock, the world's largest asset management company, then overseeing a hoard in excess of $10 trillion.

The Ford company was now organized to deliver rewards to shareholders, and to a degree that the Dodge brothers could scarcely have imagined.

The company had broken with another core tenet of its past— Henry Ford's obsession with self-sufficiency. The vertical integration that had been his lodestar had long since fallen into disfavor within the management ranks. The corporation had sold off many of its constituent pieces, including a bankrupt steel mill to a Russian firm.

Much like every multinational business, Ford had exploited globalization to lower its costs. The company was directly purchasing about 1,800 parts while relying on its suppliers to procure tens of thousands of others from their own sources scattered around the world. It was buying steering columns from Hungary, air bags from Sweden, suspension parts from Japan, door hinges from Canada, instrument panel components from China, and sunroofs from Germany. It was assembling a popular new pickup truck, the Maverick, in Mexico.

The F-150 pickups coming off the assembly line at the Rouge drew on this international supply chain. Nearly half the parts they were using had been made somewhere beyond the United States or Canada.

Ford was an exemplar of the virtues of the global supply chain, tapping the specialized attributes of factories around the world. And yet the company had left itself vulnerable to getting pinched in a way that would have left its founder aghast. It had relied overwhelmingly on suppliers who themselves depended on a single company on the other side of the Pacific to manufacture a crucial component—the computer chips that had become the brains of modern-day cars.

That single supplier, Taiwan Semiconductor Manufacturing Company, had become indispensable to the global automobile industry, producing between 80 and 90 percent of the computer chips going into cars worldwide.

During the first months of the pandemic, in early 2020, automotive companies had concluded that demand for their vehicles was about to plunge. They canceled orders for chips, prompting Taiwan Semiconductor and other major chipmakers to slash their production.

The forecasts had proved too gloomy. By the middle of 2020, people in the United States and other wealthy countries were buying cars as a way to limit the risks of encountering other humans on public transportation. Families were purchasing SUVs to go on outdoor adventures in a world suddenly devoid of indoor amusement.

By the time the automakers realized that their assumptions were wrong, it was too late. The chipmakers could not quickly respond to a sudden increase in orders. Boosting capacity required billions of dollars of investment and many months. The world found itself confronting a severe shortage of chips.

As Ford and other automakers scrambled to secure chips, they discovered that they were relatively unimportant to their suppliers. What chips were being made were being sold primarily to electronics giants like Apple, Samsung, and Sony, whose sales were multiplying as people confined to their homes desperately sought diversion from isolation and omnipresent death.

Unable to buy enough chips, the Ford company was forced to slow production. By the middle of 2021, it was temporarily halting assembly lines at eight plants in North America—from Chicago to Kansas City to Mexico.

"It's exactly the kind of thing that Henry Ford feared," said Matt Anderson, curator of transportation at the Henry Ford, a museum exploring his legacy and the history of American innovation, in his hometown of Dearborn, Michigan. "He was just so fearful of supply chain disruptions, or of having his supplies cut off."

At least part of how Ford had found itself dependent on a single, faraway supplier reflected the company's bowing to its founder's nemesis—shareholder interest. In the three years before the pandemic, the corporation had distributed dividends to shareholders totaling $7.9 billion, or 70 percent of its profits.

Ford executives pushed back on the idea that the gratification of investors had been responsible for the company running out of chips. The F-150 alone required more than eight hundred types of semiconductors, undermining any notion that Ford could have achieved self-sufficiency, they said. Chips have limited shelf lives. Ford could not have simply used the cash it distributed as dividends to amass a stockpile.

For Ford, making its own chips, or even limiting its suppliers to North America, would pose "a Herculean task that would be very asset and capital intensive, and just not realistic," the company's chief industrial platform officer, Hau Thai-Tang, told me.

But as Thai-Tang acknowledged, the parameters of what was realistic were largely determined by investors. Like the rest of the manufacturing world, Ford had long avoided stocking lots of extra components as a way to limit costs. The pandemic had revealed the consequences of taking that concept too far.

"We're certainly reflecting on the past two years," Thai-Tang said.

On the day I visited the Rouge, many of the pickup trucks coming off the line were destined not for the driveways of American families but for the parking lots in the shadow of Ford's corporate headquarters. There the vehicles sat, awaiting a shipment of chips that could bring them to life.

One lot jammed with stranded trucks sat directly across the street from a campus named for a local icon: the front windows of Henry Ford Elementary School now looked out on hundreds of vehicles waylaid by an epic failure to manage the supply chain.

Ford was far from the only company that had eschewed extra supplies as a drag on the bottom line. Most global businesses were operating with little margin for trouble.

Here was a key part of the explanation for why the halting of factories in China quickly yielded shortages of parts everywhere else, and why Walker's company found itself fretting over its ability to secure needed materials.

For decades, executives had run businesses in the thrall of an obsession with Just in Time delivery. They had concluded that stashing extra parts and products in warehouses as a hedge against trouble was a poor use of their money.

Just in Time was a sensible concept that had been mastered by Ford's formidable competitor, Toyota, one of the most successful companies in the history of commerce.

But business consultants had taken the idea and twisted it into a crude imperative to slash inventory as a way to lower costs and send share prices higher. They traded the long-term resilience of the supply chain for the short-term pleasuring of the investor class.

This reality had been building for decades. The pandemic would reveal the extent to which the world was unprepared for upheaval.

CHAPTER **3**

"NO WASTE MORE TERRIBLE THAN OVERPRODUCTION"

The Roots of Just in Time

Walker had good reason to worry about the fate of his *Sesame Street* order. The continued disruption of factories in China yielded shortages of parts. And this reality was amplified by a basic feature of contemporary manufacturing. The global economy had entered the pandemic with scant margin for error.

For decades, factories had maintained minimal volumes of components and raw materials. This was the outgrowth of the revolution that had seized the business world, the embrace of Just in Time. And that was in large part the result of the creative powers displayed by a factory manager in Japan named Taiichi Ohno, the man credited with leading Toyota out of the devastation of World War II.

Ohno was ruled by a deep revulsion for any squandered resource. Famously unpretentious, he was a creature of the factory floor who disdained meetings and the argot of technocrats, preferring blunt verbal interrogation.

The grandly named Toyota production system that defined his legacy was not the product of some master plan. Rather, it was the

sum total of a series of iterative, often tentative improvements re-
fined over decades through a process of experimentation. The company did not even start using the term until the 1970s.

Yet the impacts of Toyota's relentless process of transformation
are difficult to overstate. Its approach to manufacturing changed the
way that products are made around the globe.

At the center of Toyota's designs was Just in Time manufacturing. Rather than warehousing huge quantities of parts and raw
materials—an undertaking that consumed time, physical space, and
money—Toyota configured its plants, and even those of its suppliers, to ensure that components arrived at the assembly line just as
they were needed for the next step in making a vehicle.

As Toyota transformed itself from a miniscule player on the world
stage into the largest automaker on earth, it also claimed distinction
as the leader of a global movement that refashioned business, one
operating under the banner of so-called lean manufacturing. Management consultants peddled reports promising that if corporate
executives bought into the Toyota mindset, adopting its ways, they
would realize extraordinary gains in efficiency, lower their costs,
and increase their profits.

This movement played out far beyond auto factories, eventually
working its way into nearly every crevice of the global economy.
Lean became the concept guiding the operations of pharmaceutical companies and hospitals, meatpackers and railroads, makers
of computer chips and medical devices. This was part of the reason
why the shutdown of Chinese factories swiftly yielded substantial
shortages—because, even before the shock, the inventory had been
lean.

"We believe that the fundamental ideas of lean production are
universal—applicable anywhere by anyone," declared the authors of
an especially influential book, *The Machine That Changed the World*,
which chronicled Ohno's refashioning of Toyota as the template
for reinvigorating manufacturing worldwide. "The adoption of lean

production, as it inevitably spreads beyond the auto industry, will change everything in almost every industry—choices for consumers, the nature of work, the fortune of companies, and, ultimately, the fate of nations."

Ohno himself portrayed Toyota's philosophy as far more than an approach to manufacturing products. He offered it as a way to analyze any human undertaking, a surgical means of identifying and eradicating waste. His management philosophy resonated like a clarion call for the denizens of corporate offices. He was the bearer of the powerful truth that it was better to make too little than to make too much.

"The Toyota production system," he declared, "is not just a production system. I am confident it will reveal its strength as a management system adapted to today's era of global markets and high-level computerized information systems."

The lean movement achieved many of its aims, proving potent and enduring. Companies that embraced its principles became more productive and innovative, elevating the quality of their goods. For better and worse, the Just in Time mindset combined with advances in transportation, the rise of the internet, and the forging of international trade deals to accelerate globalization, as companies looked across oceans for new suppliers. Many of these impacts proved beneficial, opening markets to companies that produced more jobs at higher wages while delivering a bonanza of affordable goods.

Yet like many powerful notions that hold merit, Just in Time was eventually exploited in the pursuit of craven objectives. Beginning in the 1980s, consultants hijacked the concept as justification for a series of short-sighted corporate undertakings that enriched shareholders at the expense of everyone else.

As Ohno grasped clearly, even excellent concepts can be abused in the service of damaging ends.

"As in everything else," he wrote, "regardless of good intentions, an idea does not always evolve in the direction hoped for by its creator."

Ohno's quest for efficiency was shaped by the unique challenges of his archipelago nation.

Japan's topography was defined by a series of rugged mountain ranges, limiting the availability of arable land and confining the bulk of the population to narrow swaths of terrain. This geographic reality had long forced Japanese companies to economize on resources while placing a premium on designing clever ways to limit waste.

In contrast to the United States, where the frontier identity ran deep, Japan had no room for sprawl or undisciplined lunges for scale. Especially not in the late 1940s, when Ohno asserted his influence over Toyota. Having emerged defeated and ravaged from World War II, the nation's industrial capacity was decimated.

Ohno was born in 1912 in Dalian, a city in northeastern China then occupied by Japanese military forces. After graduating from a technical school in Nagoya—known as Japan's Detroit—he took a job as a manager at the Toyoda Spinning and Weaving Company, a family-owned textile business. A decade later, with the textile business flagging, he shifted to the family's fledgling automobile business.

It was 1943, and the war was still underway. Ohno's initial job entailed producing aircraft parts for Japan's military campaign. By the end of the conflict, he was based at Toyota's original factory near Nagoya, where he oversaw assembly, focused largely on making trucks. There, he used one part of the factory as his own laboratory. He cut from four to three the number of workers assigned to a single machine, while encouraging the workforce to brainstorm improvements that could increase production.

By 1950, Ohno's factory was making about a thousand trucks per month, roughly six times as many as when he began. But that burst of activity outstripped the demand for vehicles in the still-anemic

Japanese economy. As Toyota confronted a massive pile of unsold inventory, it teetered toward bankruptcy.

Unable to pay its bills, Toyota laid off some two thousand people—about one-fourth of its workforce—triggering a crippling strike that lasted three months. In exchange for labor peace, the company agreed to confer profit-sharing and lifetime employment guarantees on its remaining workers.

Toyota's pressure to maximize the productivity of its workforce stemmed from something that today seems ironic. Under its postwar Japanese occupation, the United States imposed American labor laws that restricted the company's ability to fire workers. Given its obligation to retain workers regardless of sales, Toyota's management felt extra compulsion to seek efficiency.

Toyota's near-death experience supplied Ohno the realization that formed the core of his management philosophy going forward. Mass production cloaked myriad forms of waste.

Perpetually short of capital, Toyota was in no position to manufacture all of its components, instead relying on a network of suppliers. Those suppliers produced single parts in huge batches— ten thousand at once—without coordination. Toyota's workers often found themselves waiting around for required parts before they could assemble finished vehicles.

"We desperately needed to find a way to secure the parts we needed, when they were needed, and in the amounts needed," Ohno told an interviewer decades later.

He did not come to this conclusion by himself. Rather, he executed on an understanding acquired over the previous decades by generations of Toyoda family leaders.

On a visit to the United States in 1910, the founder of Toyota's auto business, Sakichi Toyoda, had been enthralled by the sight of Ford's Model T navigating urban thoroughfares. Returning to Japan, he brought an eagerness to catch up with American industry. He was also the first person that Ohno heard utter the term "just in time manufacturing."

Forty years later, Sakichi Toyoda's nephew, Eiji Toyoda, undertook his own visit to the United States, then in the midst of its postwar economic boom. He made pilgrimage to Ford's Rouge plant. There, he was mesmerized by the colossal scale of the operation, yet certain that it could not be duplicated in Japan. Ford was running as many as three hundred individual production lines—a series of workstations dedicated to completing one element of building a car. Toyota lacked the cash to contemplate such scope.

Eiji Toyoda—who would later serve as the company's president—also studied another realm that proved crucial in the genesis of Toyota's production model: the American supermarket. The shelves of a supermarket seemed a natural representation of Just in Time management. Items had to be stocked in sufficient quantity to ensure that customers could buy what they needed, yet not in such abundance that they exceeded the available display space or went bad before they were sold.

This, Toyoda realized, was the same principle that should guide a factory's handling of inventory.

On his own travels to the United States in the mid-1950s, Ohno's observations of supermarkets yielded an insight that became a crucial dimension of the Toyota production system. At each phase of the factory operation, the company should make only enough of any given item—a part, a component, a finished product—to replenish what had been used up further down the line.

He contemplated the milk on the supermarket shelves. Stores stocked only enough to replace what was purchased—not less, or some customers would leave empty-handed; not more, or managers would be forced to throw away spoiled cartons. How much milk customers bought determined the volume that distributors should deliver from surrounding dairy farms. That volume in turn influenced how many cows those dairies had to maintain, how much cattle feed local farmers should grow, and how much fertilizer local distributors needed to keep on hand.

This, Ohno reasoned, was the way that Toyota's factories should operate.

His revelation may now seem so commonsensical as to be self-evident. Yet in embracing it as the corpus of his management philosophy, Ohno was transcending the conventional wisdom of the era.

Under the model pioneered by Ford, factories scrambled to make as many cars as possible. Suppliers rushed to build up inventory for fear of being caught short. The result was not infrequently a glut of unsold product.

"There is no waste in business more terrible than overproduction," Ohno declared.

Ohno's analysis became the impetus for a tightly controlled system of managing inventory centered on kanban—written manufacturing instructions that accompanied every batch of parts produced within Toyota's operations. The workstation on the assembly line that installed car doors would take possession of fifty doors pressed by the unit that made them and would hand over kanban that would trigger the production of fifty more. In this fashion, Toyota ensured that it was always making the right volume of parts and finished vehicles—no more, no less.

By the 1960s, kanban were in use throughout most of Toyota's factories. The company soon spread the system to its outside suppliers, sending back empty boxes used to ship parts as an indication to make more.

The oil shocks of the early 1970s sent gasoline prices soaring, propelling Toyota's expansion. Lighter, fuel-efficient vehicles had always made sense in Japan—a major oil importer. Now, they were in demand from London to Los Angeles. By the early 1980s, Japanese automakers had claimed nearly 30 percent of the market for cars worldwide, up from less than 4 percent two decades earlier.

Toyota was by then instructing American automakers on the finer points of its system, having set up a joint venture with General Motors at a formerly shuttered factory in the San Francisco Bay Area. Ford was dispatching its executives to Japan to visit Toyota's factories in a reversal of the path that had brought the Toyoda family to Michigan.

Ohno's book—first published in Japanese in 1978—was released in English a decade later, and gained currency among business consultants. By then, two-fifths of the largest manufacturers in the United States had embraced some version of Just in Time.

In 1990, a trio of researchers at the Massachusetts Institute of Technology published *The Machine That Changed the World*, their exegesis of Toyota's production model, while coining a new catchphrase to describe the emerging global movement—lean manufacturing. That volume sold an astonishing six hundred thousand copies while handing consultants a bible in their holy mission to cut costs.

Ohno had detailed a comprehensive approach to manufacturing, one that included a rigorous dedication to quality and an understanding of the value of shared purpose among the workforce. The production system that he had overseen guided a relentless process of investment and expansion, turning Toyota into an enterprise with factories in North America, Europe, Southeast Asia, and China, while employing 370,000 people worldwide.

Yet in presentations to the people running multinational companies, Ohno's philosophy was easily distilled down to an imperative to eschew inventories as a capital-wasting threat to the bottom line. The rest of his message was lost amid a special focus on the profit-enhancing powers of Just in Time.

"Our ancestors grew rice for subsistence and stored it in preparation for times of natural disaster," Ohno wrote. "A person in business may feel uneasy about survival in this competitive society without keeping some inventories of raw materials, work-in-process, and products. This type of hoarding, however, is no longer practical. Industrial society must develop the courage, or rather the common sense, to procure only what is needed when it is needed and in the amount needed. This requires a revolution in consciousness, a change of attitude and viewpoint by businesspeople."

That change was on the way, though not as Ohno had intended.

CHAPTER 4

"THE LEAN TALIBAN"

How the Consulting Class Hijacked Just in Time

The cult of business consulting seized on Just in Time as the central tenet of a new faith in the wisdom of cutting costs. It took Toyota's commonsensical approach to running factories and reconfigured it into a religious dictate to unload inventory, the essential commandment in a holy mission to liberate corporations from the tyrannical constraints of tradition.

The extremists in this campaign helped create the world of recent decades, one in which multinational companies have made themselves vulnerable to trouble down the road while burnishing their balance sheets right now.

That trade-off made little sense over the long haul. It left the supply chain exposed to the very sort of upheaval that was giving Walker fits over his inability to secure the materials needed to make his *Sesame Street* order. But the logic was compelling for the select group who stood to benefit: the shareholders of publicly traded corporations.

Knut Alicke was among the apostles in this denomination, an evangelist in the service of Just in Time.

Alicke works for McKinsey & Company, a business consultancy whose very name is synonymous with power and influence. In its near century of existence, the firm has advised the world's most prominent companies, from the masters of finance to technology giants. It has counseled the Saudi monarchy, Russian conglomerates linked to Vladimir Putin, and Chinese state-owned companies that answer to the Communist Party. It has ministered to key pieces of the American national security apparatus, including the Central Intelligence Agency.

As one writer put it: "If God were to remake the world, he would call upon McKinsey for assistance."

Over the decades, McKinsey has cultivated a reputation as a ruthless diviner of complex truths, its insights gleaned from a rigorous, data-driven process of inquiry. It has recruited graduates from the world's most prestigious campuses, indoctrinated them into the McKinsey way, and inserted them into the inner sanctum of corporate power.

The firm has prospered above all by catering to the most visceral impulses of the people running major businesses—in particular, their insatiable demand for new ways to lift their stock prices.

Alicke joined McKinsey in 2004, after gaining a doctorate in mechanical engineering at the Karlsruhe Institute of Technology in Germany. Soft-spoken and accommodating, he projects a nerdy sense of authority in the engineering realm, along with the patience to break it down for mere mortals.

For most of his years at the firm, he has pressed the mantra that the key to vitality is embracing the concepts of lean manufacturing. He first absorbed that school of thought while reading *The Machine That Changed the World,* the book that proclaimed the power of the Toyota production system.

Alicke's first project at McKinsey entailed streamlining warehouse operations for a German maker of soap and skin cream. He later counseled automakers on how to reduce inventory. But he has also proselytized for *lean* in industries far removed from the obvious targets.

In 2010, he coauthored a report urging pharmaceutical companies to dramatically reduce their warehouse costs, in part by eliminating job security for their workers. He urged companies to downgrade full-time jobs into "flexible" positions, meaning part-time stints in which employees were effectively on call at the whims of their bosses.

"By reducing the notice period for shift schedules to one or two days, facilities can more closely match on-site staffing to demand, raising efficiency by up to 15 percent," his report declared.

Even greater gains were waiting to be harvested by resorting to a "super-flex temporary workforce" that could be dispatched to warehouses via text messages within "only a few hours' notice, allowing the facility to respond on the same day to unexpected demand peaks."

In this usage, *going lean* had become a euphemism for transferring control and economic security from ordinary workers to the executive ranks. Companies could lower their labor costs by forcing employees to make themselves available at all times, eliminating their ability to attend to the needs of their families while removing certainty over their household budgets.

In the business vernacular, *lean* and *flexible* had become components of the same basic idea: that huge companies stocked with full-time workforces were fatally constrained in their ability to innovate. They were slow and bumbling, bureaucratic and ponderous, as opposed to nimble and agile.

A flexible workforce could be quickly redirected toward whatever enterprise seemed most lucrative at any given time. It could also be dismantled on a whim, the workers cut loose as needed, without having to worry about severance payments, union grievances, or other hassles that conflicted with the bottom line.

Like most McKinsey initiatives, the firm had promoted the merits of the flexible workforce with a veneer of dispassionate science.

Back in the early 1980s, two directors at McKinsey had undertaken a survey of their most successful clients, seeking to identify practices worthy of emulation. The bestselling book that resulted from their project, *In Search of Excellence,* laid out the approved course for any company grappling with the complexities of globalization and digitalization: fire large numbers of people and replace them with temporary workers who could be terminated anytime. This became the template for the era.

Toyota's success had spurred the Just in Time revolution. McKinsey had pushed that idea beyond the management of parts and warehouses, extending it to the oversight of human beings. People were inventory, too. The more workers a company had on hand, the higher its costs and the greater the barriers to reconfiguring its business to exploit new opportunities.

The legions of temporary workers who, over recent decades, have displaced full-time positions in major economies around the world—from the United States to France to South Korea—represent the triumph of this thinking. Encouraged by consultants, the executive class was adopting the view that their workers were less a source of goods and services than a drag on the balance sheet, one best contained with remorseless dedication.

Still, Alicke and his colleagues counseled clients to "foster a mindset of continuous improvement" and safeguard "staff satisfaction" by employing the usual corporate fare, like recognizing an employee of the month. For workers, the unceasing torment of not knowing if they could take a child to the doctor would be compensated by their name tacked on a noticeboard adorned with gold stars.

Alicke expressly urged companies engaged in selling consumer goods to set aside concerns about maintaining a stash of extra inventory as insurance against unforeseen troubles. In the age of the Web and international transportation networks, anything that was in short supply could be swiftly ordered and delivered. Sophisticated computer programs could sift through vast troves of data to anticipate changes in demand with extraordinary precision, rendering large warehouses an expensive anachronism.

"New planning algorithms will significantly reduce the uncertainty," Alicke and his coauthors wrote, "making safety stock unnecessary." By infusing their businesses with lean management, a company could achieve "an overall inventory reduction of 50 to 80 percent."

Above all, Alicke described a world in which the supply chain was not just a way to move products from factories to customers, but a means of fattening their profits.

"This is the opportunity for the supply chain organization to rise from being an execution-focused function to being at the center of growth," he wrote in 2018.

The merits of this proposition were seemingly affirmed by the fact that the world's most successful companies were embracing *lean*. Apple had long been ruled by a singular dedication to Just in Time. Its chief executive, Tim Cook, once described inventory as "fundamentally evil."

"You kind of want to manage it like you're in the dairy business," Cook said, channeling his inner Toyota. "If it gets past its freshness date, you have a problem."

But as COVID-19 shut the Chinese factories that assembled the iPhone, Cook faced a different problem. His company, along with countless others, suddenly could not purchase the parts needed to make his company's products. And if Apple could not find the components it needed, what hope was there for a small start-up confined to Mississippi?

Until the pandemic, Tim Cook had not been wrong. Cutting inventory really had tended to enrich corporations. That fact had been validated for years on stock markets from Hong Kong to London to New York.

Ever since the 1970s, executives of publicly traded companies had been consumed with increasing the value of their shares—or risk getting fired. And they personally pocketed some of the gains, because their pay packages were increasingly dominated by grants of shares.

The management playbook included myriad ways for a company

to lift the value of its shares, from firing workers—a Wall Street favorite—to raising prices or increasing sales.

But one key metric stood out as especially vital among professional money managers: the so-called return on assets. This was the amount of profit that a company generated as a share of the value of all the things that it owned. Companies that succeeded in increasing their return on assets made their shares more attractive to investors.

From a pure accounting standpoint, cutting inventory immediately enhanced this important data point, because it reduced the total value of a company's assets. This was straightforward math. Less inventory meant that the company's profits were divided by a smaller number, yielding a higher ratio.

"To the extent you can keep reducing inventory, your books look good," said ManMohan S. Sodhi, a supply chain expert at the City, University of London Business School.

This basic arithmetic exercise proved irresistible to publicly traded corporations. Between 1981 and 2000, American companies reduced their inventories by about 2 percent every year. And by 2014, American companies were holding roughly $1.2 trillion less in inventory than they had in the 1980s.

This success translated into money for McKinsey. As Walt Bogdanich and Michael Forsythe revealed in their bestselling book, *When McKinsey Comes to Town*, the firm's compensation was sometimes tied directly to the cost savings that it yielded for clients. That gave consultants a direct incentive to cut spending—on labor, on inventory, on equipment, on everything.

Just as Henry Ford had warned, the primacy of the investor class was transforming companies from makers of goods and providers of services into entities organized around gratifying shareholders, with their stocks turned into the product itself.

At the companies that McKinsey advised, Alicke generally met with operations people who attended to the everyday realities of the factory floor. They were intimately familiar with the complexities of production and open to the merits of the lean mindset—not just cutting inventory but also eliminating choke points and reducing defects.

But the operations people reported to executives who were narrowly obsessed with financial considerations. Somewhere between the factory floor and the C-suites, the breadth of the lean conversation was condensed into a simple dictate to empty warehouses.

"Up the hierarchy in finance, they just see that 'lean' means low inventory and they say, 'Let's push it,'" Alicke told me.

Inside McKinsey, the consultants who were most zealous about slashing inventory were sometimes referred to as "the Lean Taliban," given their eagerness to eradicate those who betrayed less than fanatical devotion to the cause.

From his perch at a factory in Minnesota, Jerome Bodmer watched the insurgency storm the gates. He was working in the purchasing department of a company called Onan, which manufactured industrial generators just up the Mississippi River from Minneapolis. Its customers included large hospitals and real estate developers, who required absolute certainty that their backup supplies of electricity would work in the event of power failures.

Onan's largest generators were so hefty that they required installation by crane, which meant that they had to be delivered right on schedule. This argued for maintaining an ample supply of parts and components, ensuring that the factory could always complete its orders. But as a group of consultants showed up at the factory one day in the mid-1980s, Bodmer was astonished to hear them prescribe the opposite approach.

The team from McKinsey's offices in Chicago assembled in a conference room—one older man accompanied by a half dozen junior associates, all of them donning slick suits, most of them fresh from Ivy League universities. They insisted that the company had to go lean.

"They told us, 'We can free up assets by reducing your inventory,'" Bodmer recalled. "'We are here to help you figure out how.'"

Bodmer was then in his early forties. He had a business degree from the University of Minnesota, plus two decades of experience working in local manufacturing. McKinsey's advice appeared to collide with common sense. What would happen when, inevitably, some supplier ran into trouble and failed to deliver? Didn't it make more sense to keep extra parts on hand?

The McKinsey people assured him that he had no idea how the modern world worked.

"You were definitely made to feel like you were a lesser individual, and they were the ones who knew, and you were the ones that would fall in line and get the work done," Bodmer said. "This is Minnesota. We were Minnesota nice. But in the back of your mind, you're like, 'How do you guys know all this when you haven't done anything yet?'"

When some of his colleagues pushed back, sharing their fears about excessive reductions in inventory, the McKinsey team ratted them out to Onan's senior executives.

"The consultants would bring that back to the leadership group and say, 'Your people aren't cooperating,'" Bodmer said.

At first, McKinsey focused on reducing expensive items, like industrial engines imported from Japan. But once they exhausted those targets, they moved on to cheaper goods like metal brackets and fasteners. For years after, Onan frequently ran out of basic parts, delaying its orders by weeks.

"We'd be waiting to finish a generator that we could sell for a quarter-million dollars, because we didn't have a $5 sheet metal bracket," Bodmer said. "The focus on return on assets was really intense. That caused some really stupid decisions."

In the early 1990s, Onan was purchased by Cummins, a publicly traded company that manufactured machinery used to generate electricity. That deal intensified the pressure to satisfy Wall Street's cravings for enhancements to the return on assets, Bodmer said.

Invariably, near the end of the month, some big-ticket item like an engine would arrive, imperiling that metric by increasing the inventory on the books. Managers would refuse to take delivery, instead placing incoming purchases on trailers and stashing them outside in a parking lot until the beginning of the following month. Only then would the factory sign off on accepting new inventory, beginning the cycle anew.

Delays in accepting parts often held up the factory's production. That forced the company to pay extra to send finished orders by air freight instead of trucking in order to appease customers.

In short, the company was willing to waste real money on speedier shipping to correct self-imposed delays that were the cost of burnishing the optics of its most recent quarter.

"This would happen every month," Bodmer said. "It was always a focus on short-term results."

The strategy that McKinsey was prescribing for its clients seemed, on its face, ridiculous. Cutting inventory to the bone provided instant gratification for shareholders, yet at the cost of the company's ability to serve customers. Cummins was losing sales, damaging its reputation, and undermining morale inside the factory.

But the calculus made perfect sense when viewed from the perspective of the client that was actually engaging McKinsey—not the company and its workers, but the executives who were answerable to investors.

For corporate executives, cutting inventory freed up cash for other purposes. They could use it to hand themselves extravagant pay packages as a reward for their strategic brilliance in going lean. They could distribute the money to shareholders in the form of dividends, or they could deploy it to purchase shares of the company's own stock—a move that tended to push share prices higher.

Once upon a time, companies were essentially barred from buying back their own shares, a course construed by market watchdogs as an illegal form of stock price manipulation. But then the Reagan administration ushered in a period of fervent deregulation. The financial services lobby soon liberated Wall Street. Over the decades, buying back shares became standard practice for publicly traded companies, an ordinary means of pleasing investors.

In the ten years leading up to the pandemic, American companies devoted more than $6 trillion—an amount in excess of the annual economic output of Japan—to finance purchases of their own shares, roughly tripling their previous pace of buybacks.

A similar story played out in major economies worldwide. Companies in Japan, Britain, France, Canada, and China increased share buybacks fourfold during these years, though their total purchases were a mere fraction of their American counterparts'.

Share buybacks and dividends generally worked as intended. They lifted share prices, adding to the wealth of investors. But the gains came at the expense of whatever else management might have done with its cash, such as buying new equipment, hiring workers, introducing new products, or salting away extra parts as a hedge against disruption.

By the time COVID-19 arrived, the dangers of finance taking precedence over prudence had been evident for years, even as they were blithely ignored by corporate chieftains. Executives were too busy cashing their stock options to worry about killjoy distractions like preparing for some hypothetical natural disaster that probably would not happen in the next quarter.

More than two decades before the pandemic, in 1999, an earthquake registering 7.3 on the Richter scale rattled Taiwan, by then a leading center of computer chip manufacture. The quake shut down a major industrial park where key components for chips were made, generating shortages around the globe. By the time the Christmas season unfolded three months later, parents were struggling to locate popular products like Nintendo's latest gaming console. Computer makers suffered hits to their sales as a shortage of chips limited their production.

By 2005, some of McKinsey's own people at its offices in Tokyo were sounding the alarm that Just in Time had been carried to perilous extremes. In an internal presentation titled "When to Stop Dieting," consultants concluded that some factories had cut inventories so deeply that they had jeopardized their ability to make products without interruption.

At workstations inside some Japanese auto factories, boxes with enough room to hold eight hours' worth of parts were stocked with only a two-hour supply—a decision informed purely by an obsession with the numbers disclosed in reports. With no change to the physical layout, the same boxes could have held four times as much, reducing the frequency of replenishment. Meanwhile, workers were sometimes forced to halt production while they waited for the next infusion of parts.

That same year, the journalist Barry C. Lynn presciently described the mounting dangers of excessive reliance on Just in Time.

"Our corporations have built a global production system that is so complex, and geared so tightly and leveraged so finely, that a breakdown anywhere increasingly means a breakdown everywhere," Lynn wrote. "Our corporations have built the most efficient system of production the world has ever seen, perfectly calibrated to a world in which nothing bad ever happens. But that is not the world we live in. Not only is human civilization riven routinely by earthquakes and hurricanes, but so too it is shattered by wars and acts of terror and simple human error. Which means it is only a matter of time until we experience our next industrial crash, perhaps one much worse than any we have yet known."

Those words were wise. They were also ignored. Executives running companies had no incentive to spend money in pursuit of nebulous concepts like resilience. They were rewarded for cutting costs, a pursuit that they justified in the name of gratifying consumers with lower prices, even as they handed most of the bounty to shareholders.

Mike Ullman could attest to the risks of failing to satisfy investor demands for going lean. In the early 2000s, as the chief executive officer of the once iconic American retailer J.C. Penney, he reduced inventory to avoid getting caught with unsold clothing. Too often, his company had been forced to unload mountains of leftover apparel at steeply discounted prices—a hit to the bottom line. By keeping less inventory, J.C. Penney would be able to charge more for its goods.

But Ullman did not act aggressively enough to satisfy Bill Ackman, a hedge fund manager and so-called activist investor, who captured stakes in publicly traded companies and then forced change. In October 2009, Ackman and another pirate on the high seas of finance, Steven Roth, secured a 26 percent stake in J.C. Penney. By the following year, they had exiled Ullman while replacing him with their chosen recruit, a veteran of Apple named Ron Johnson. They clearly expected their new leader to infuse J.C. Penney with Apple's approach to business, including the ruthless management of inventory.

"Ron Johnson is the Steve Jobs of the retail industry," Ackman declared.

That episode ended in a spectacular fiasco. Johnson slashed inventory to free up $500 million in cash, burnishing the company's balance sheet, while allowing it to pay out dividends. But he overdid it, tormenting customers who had traditionally flocked to J.C. Penney for its well-advertised sales. As the company's stock price plummeted, Ackman turned on Johnson. The new boss was out after less than a year and a half in the job, replaced by his predecessor.

Still, Johnson's demise merely illustrated the ultimate dictate in a world ruled by financial concerns: never allow the stock price to fall. Even as Ullman was restored to the corner office, the reasons behind his initial banishment reflected incentives that had in no way been diminished. Lean remained the mantra.

In March 2011, Japan was devastated by an earthquake registering a terrifying 9.0 on the Richter scale, followed by a typhoon that overwhelmed the northeast coast of the main island of Honshu. The wall of water swamped a nuclear power plant in the province of Fukushima. Nearly twenty thousand people lost their lives. Millions of households and businesses lost power for months, shutting key swaths of Japanese industry. The disaster zone included a single plant tucked into a hillside that by itself produced about one-fifth of the world's advanced silicon wafers, a key element used to make computer chips.

The impacts of the disruption swiftly rippled out around the globe. From Germany to Spain to North America, auto plants were forced to slow or halt production in the face of a shortage of electronic sensors. The French automaker Peugeot-Citroën disclosed that it routinely stocked only ten days' worth of an air-flow sensor that was suddenly in critically short supply.

Once again, experts seized on the episode as an indication of an excessive reliance on Just in Time, one that had left the global economy exposed.

"There should have been fail-safe measures," said Kenneth Grossberg, a business professor at Waseda University in Tokyo. "How could you make yourself so vulnerable?"

Leaders of major companies promised to use the tragedy as a learning experience, while making amends.

"We'll do a retrospective on what worked best and what didn't, and how to change things to make our supply chain more resilient," a senior executive at Hewlett-Packard, the computer giant, told my colleague, Steve Lohr.

But once again, in the years after the Fukushima disaster, the commercial world reverted back to how business had been handled before, disdaining extra inventory as a threat to the bottom line. Japan might have been shaken, but the incentives that guided multinational companies remained intact.

Some experts used Fukushima to explore the growing perils of global integration. In a book published in 2014, *The Butterfly Defect*, a pair of academics, Ian Goldin and Mike Mariathasan, warned that the movement of people, goods, and capital around the globe—a force they celebrated as progress—was increasingly a vector for the spread of systemic risks. The global financial system, the internet, security networks, and the supply chain were all susceptible to threats more easily spread by interconnectedness.

Their book included a full chapter on the special risks posed by pandemics.

"There is a significant probability that a pandemic will strike a financial center such as New York or London and, through disease, quarantine, panic, or the collapse of secondary services (transport, energy, information technology, or other), lead to at least a temporary isolation of major players in the global system," they wrote. "Many experts believe that it is a question of when, not whether such events will occur."

The supply chain was uniquely vulnerable to a pandemic, they noted while emphasizing the consequences of extreme reliance on Just in Time. This, they concluded, was the result of investor interests having seized primacy over resilience. The executives running corporations were more afraid of Wall Street than the ravages of some hypothetical disaster.

Here was the fundamental problem—a dynamic still in force today. A corporate chieftain who slashed inventories while handing cash to shareholders was rewarded with the wherewithal to wake up on a private yacht. An executive who used that money to warehouse parts as insurance against some hypothetical calamity was invited to surrender their office.

Between 2009 and 2018, the companies that made up the S&P 500 index, a broad gauge of the stock market, spent $4.3 trillion—more than half their total profits—on purchases of their own shares, plus another $3.3 trillion on dividends. All told, these large companies distributed more than 90 percent of their profits to shareholders via buybacks and dividends.

Over those same years, American manufacturers reduced their inventories, further weakening their ability to weather whatever storm lay ahead.

Those two trends—a surge of buybacks, and the depletion of inventory—were directly related.

Executives had learned from the Taiwan earthquake and then from Fukushima that it paid to forget about such unpleasantness as quickly as possible. Maintaining lean operations allowed the dangers to build to the levels exposed by the shock of the pandemic. Yet in the intervening years, the Lean Taliban sent stock prices higher, which added zeroes to the net worth of the people who hired McKinsey.

"After Fukushima, there was a six-month rethink," Alicke told me. Corporate executives who had previously preached the gospel of Just in Time added a new phrase to their lexicon: Just in Case. This compelled some to consider holding greater inventories of parts.

"Then it stabilized," Alicke added. "And then people were back into, 'Hey, why don't we have very low inventories.'"

By the spring of 2021, people like Alicke had a lot to explain. Around the globe, patients stricken with COVID-19 were dying for a lack of ventilators. Hospitals could not secure antibiotics.

Children were resuming classes without enough school supplies. And the auto industry remained crippled by a shortage of chips.

I called Alicke to ask about the culpability of Just in Time.

"We went way too far," he told me. "There is, by design, not a lot of flexibility in the system."

The result was a reduction in the world's productive capacity so substantial that factories from Taiwan to Japan to the United States could not easily adjust to a disturbance. Even Toyota had been forced to cease production at some of its factories.

Chipmakers had shifted their production from the United States decades earlier, leaping the Pacific to Japan, South Korea, and Taiwan, in large part as a way for major users of chips like Apple to limit their inventory while tapping lower-cost suppliers. With shortages now laying waste to balance sheets and concerns about the geopolitical risks weighing on Taiwan, Intel, the American computer chipmaker, had outlined plans to spend some $20 billion to erect new factories in Arizona.

But why had the company waited so long?

At least part of the answer could be found in the $26 billion that

Intel had spent buying back its shares in the two years before the pandemic—money that might have been devoted to expanding its capacity.

The chip shortage was wreaking havoc in industries far removed from cars and electronic gadgets, withholding relief for people suffering life-threatening disorders.

In San Diego, a forty-four-year-old restaurant worker named Joseph Norwood had been waiting more than six months to receive a breathing device that had been prescribed to limit the risk of his sudden death from sleep apnea.

The maker of the devices, a company called ResMed, was producing less than one-fourth of the devices on its order books, including ventilators used by premature infants. It could not get its hands on the needed chips. Its chief executive officer, Michael Farrell, was spending much of his time begging giant chip companies to prioritize his relatively tiny industry—medical devices—over their largest customers. This campaign had yet to produce any change.

"Medical devices are getting starved here," Farrell told me. "Do we need one more cell phone? One more electric car? One more cloud-connected refrigerator? Or do we need one more ventilator that gives the gift of breath to somebody?"

For Farrell, the exercise of begging for products had brought home both the complexity of the computer chip supply chain and its overwhelming reliance on Taiwan.

His direct chip supplier could not satisfy his demand because, five levels up the chain, a maker of silicon wafers in Taiwan had exhausted its inventory. Because that plant could not deliver extra products, the next link in the chain—a company that combined the wafers with circuitry—could not boost its own production. Which meant that still another supplier—a company that bought these components and packaged them into clusters—was unable to make more of its wares.

All of which meant that ResMed's supplier of circuit boards could not buy enough needed clusters, leaving the company's factories in Singapore, Sydney, and Atlanta short of crucial elements.

So Norwood did not have his breathing device. He sat at home in San Diego, unable to work and dependent on disability payments, all the while fearful of blacking out and perhaps dying.

Here was the result of lean extremism unleashed on the chip industry.

"It's so unfortunate how money controls everything." Norwood said. "Our priorities are really skewed."

As Sodhi, the London business school expert, put it to me: "When you need a ventilator, you need a ventilator. You can't say, 'Well, my stock price is high.'"

Alicke surveyed the damage and had come around to the view that resilience had to be the new guiding light for supply chains—the next mantra guiding McKinsey's never-ending mission. It was not that business leaders had erred in entrusting their operations to consultants. It was that the consultants had to refresh their talking points.

For decades, companies had behaved as if the risks of chaos in the global supply chain were negligible. They had ceded control of their inventory to algorithms and the shipping industry in willful disregard of what might go wrong. They had set aside little to nothing as a hedge against natural disasters, pandemics, and geopolitical eruptions, to say nothing of climate change.

Those days were over, Alicke proclaimed.

"The way that inventory is evaluated will change after the crisis," he told me.

That seemed at best debatable. The incentives at work in the global economy remained unchanged. There was still no reward for executives who devoted money on bolstering their supply chains against future risk. There was still the threat of termination for those who failed to boost their return on assets.

That was true, Alicke acknowledged. He characterized his optimism as more aspirational than certain.

"It feels like the companies," he said, "they tend to forget very fast."

CHAPTER 5

"EVERYBODY WANTS EVERYTHING."

The Epic Miscalculation of Global Business

Hagan Walker was not running a publicly traded corporation, so he had no stock price to worry about. There was no upside to juicing his return on assets through creative accounting, little reason to fetishize going lean. He had a real business to run. He needed actual products made and delivered to his customers in time for the holiday season.

And still, the assumptions of the Just in Time era subtly colored his calculations.

As a start-up with limited capital, Glo had to be careful with its cash. Walker was, by nature, a conservative overseer of his company's finances. In placing orders with the factories that made his products, he tended to buy the minimum necessary to satisfy his customers.

He had faith that he could always order more whenever his stocks ran out, locating a factory in China that could quickly make what he needed. He took it as a given that space on a container ship could be easily booked at a price that seemed inconsequential. To imagine otherwise was like worrying that McDonald's might run out of french fries.

But this time, in late 2020, as Walker made arrangements for the most important order in the history of his company, his confidence in the supreme reliability of the supply chain was about to be tested.

Fortunately, he gained the counsel of a man named Calvin Zheng. He oversaw operations at Platform88, the company in China that was manufacturing his *Sesame Street* figurines.

Zheng delivered a grim report. The Chinese economy was awash in dysfunction with no certainty about when the crisis would end. Nearly every factory was struggling to find enough workers given COVID-19 quarantines and widespread fear of the pandemic. Parts and raw materials were extremely difficult to secure because so many industrial zones had been locked down, depleting stocks.

Walker was particularly concerned to learn that Platform88 might struggle to locate enough of one key ingredient: a specialty form of plastic.

His cubes required a precise combination of malleability and strength. They had to be soft enough to ensure that a kid who tried to bite the corner would not be hurt, yet strong enough to withstand the same child trying to poke their fingers through the plastic. And they had to be translucent enough for light to pass through.

Thermoplastic did the trick. But that ingredient was tough to find, given that it was used in all sorts of other products, from smartphone cases to thermoses. The Chinese government was prudently limiting the availability of plastics for relatively trifling concerns like Elmo dolls while prioritizing their use for vital goods like medical devices. Every type of plastic was soaring in price. This was an inarguably wise public health policy. Still, it left Walker with a problem.

Zheng's own factory, in the coastal city of Ningbo, was critically short of workers. Before the pandemic, some three hundred people had labored there. But after they went home to see their families during the Lunar New Year in early 2020, only about half returned. Some stayed in their villages and started small businesses. Others found factory jobs closer to home.

Zheng fretted about the availability of energy. Provincial authorities in Zhejiang province had taken to rationing electricity, fearful that volatile fluctuations in demand would take down the grid. Here was another impediment to production.

All of this argued for Glo purchasing everything it might conceivably need over the course of the year in one massive order. As Zheng kept saying, it was better to order too much than not enough, because the prices for materials were almost certain to climb, meaning that the next order would be costlier.

That warning rested on the sensible assumption that China would be inundated with more factory orders than it could manage, slowing the gears of the global supply chain.

And that understanding was based on the fact that around the world, major businesses had compounded their excessive reliance on Just in Time with a colossal failure to anticipate the impacts of the pandemic itself. They had concluded—incorrectly—that the global spread of a lethal coronavirus would destroy demand for factory goods. And once they figured out that theory was wrong, they sought to adjust by sending a flood of new orders to the factories of China.

Part of the miscalculation by the people running the global supply chain reflected a basic failure to anticipate the course of the pandemic. As COVID-19 emerged from Wuhan, the same attributes that had turned China into the center of global manufacturing became vectors for the spread of the coronavirus.

The migrant workers from rural villages arrived in factory towns where they crammed into poorly ventilated dormitories, sharing communal meals and fetid bathrooms—an ideal recipe for contamination. International air routes linking major Chinese cities with Tokyo, Dubai, Frankfurt, and other travel hubs became pathways for

COVID-19. Engineers, marketing staff, and buyer's agents traveled to factories in Ningbo and Shenzhen from corporate headquarters in Seattle and Sao Paulo, and returned home as carriers.

By late February 2020, COVID-19 was spreading death across northern Italy, the center of the nation's most formidable manufacturing operations. Then the wave moved across the rest of Europe and on to the other side of the Atlantic, to North America and Latin America. India imposed a twenty-one-day lockdown in late March. By then, the United States was leading the world in confirmed cases of the coronavirus, with more than eighty thousand.

The public health catastrophe quickly became an economic calamity as well. As governments imposed lockdowns to halt the spread of the virus, whole industries ground to a halt: travel, retail, hospitality, manufacturing, and construction. Restaurants, shopping malls, gyms, hair salons, spas, and entertainment venues devolved from destinations for pleasure to breeding grounds for a lethal plague.

Any location where humans could encounter other humans was a location to be avoided. Which meant that the people who worked in such places were no longer needed.

In April 2020 alone, more than 20.5 million Americans lost their jobs as the unemployment rate spiked to 14.7 percent, the most profound economic trauma since the Great Depression.

In Europe, joblessness rose by much smaller magnitudes, as governments stepped in to subsidize wages at companies that agreed to retain their workers. But economic anxiety spread nonetheless, and businesses suffered.

As schools shut from London to Los Angeles, parents struggled to work from home, if they were lucky enough to have jobs that allowed business to carry on via videoconferencing. The less fortunate scrambled to line up childcare for kids stuck at home while venturing into danger zones like meatpacking plants and warehouses.

Based on centuries of experience, the commercial realm had a basic understanding of the narrative arc of this story. Here was a global downturn, a worldwide recession. As companies laid off hundreds of millions of workers, they were taking paychecks out of economies. People would presumably have less money to spend, reducing demand for nearly everything. This turned out to be wrong—wildly so—but for a time the world economy hewed to that path.

Between April and June of 2020, the American economy plunged by an astonishing 32.9 percent, the most devastating downturn on record.

"This is a crisis like no other," declared the International Monetary Fund. It projected that the global economy would contract by more than 3 percent for all of 2020 and remain weak into the following year, a hit to collective fortunes reaching $9 trillion.

The people running major corporations took these bleak expectations as their cue to slash orders for the parts and raw materials needed to make their products. From northern Italy to the American Midwest to southern China, factories saw sharp reductions in demand.

The automakers were not alone in scrapping orders for key components. Major apparel brands from the Gap to Walmart to J.C. Penney reacted to the sudden abandonment of shopping malls as reason to cut purchases of products from factories in Asia.

Businesses broadly trimmed orders for new machinery, metals, and electronics components. Apple's suppliers across Asia prepared for an event once unimaginable—a drop in demand for the newest iPhones—as the company cut orders for parts.

And the shipping industry pulled back. Carriers canceled hundreds of routes on the logical assumption that weakening demand for goods spelled less need for shipping containers crossing oceans.

In April 2020 alone, container shipping carriers canceled seventy scheduled runs from Asia to the West Coast of the United States and Canada, while scrapping forty to the East Coast. By June, the largest carriers scrapped another 126 scheduled sailings between Asia and North America, and nearly one hundred routes connecting Asia to Europe.

They stashed idled ships in the warm waters of Southeast Asia while handing some leased vessels back to their owners in Greece. Cargo and cruise ship operators sold off aging ships to recycling operations, consigning them to marine graveyards in India and Pakistan, where workers demolished them to harvest steel.

By the middle of 2020, the ocean carriers had canceled between 15 and 30 percent of scheduled service on their busiest routes. They had taken so many ships out of commission that they had reduced their global freight-carrying capacity by more than 10 percent.

At the container ports in Southern California—the dominant gateway for American imports from Asia—activity slowed dramatically.

"It's very quiet," declared the director of the Port of Los Angeles, Gene Seroka, whose office looked out on the docks. The dozen ships that he usually saw in various states of loading and unloading had been reduced to four.

Then came a powerful surprise. Demand for factory goods came raging back.

Americans confined to their homes had stopped spending on steakhouses, trips to Disney World, and spa services. They were no longer wandering the aisles of department stores, cramming into concert halls, or rubbing shoulders at sporting events. But they were redirecting the money once spent on such experiences to fill their homes with goods.

With bedrooms and dining rooms suddenly doubling as offices and classrooms, Americans mass-ordered desk chairs, computers, and printers. As parents struggled to engage children cooped up with distance learning, they snapped up iPads, gaming consoles, and air hockey tables. They bought art supplies, board games, and volcano simulators.

People could not go to gyms, so they added exercise bikes to their basements. With movie theaters off-limits, Americans purchased flat-screen televisions and streaming video players, turning their living rooms into home theaters. They swapped restaurant meals for home smokers, barbecues, and pastry blenders. They bought gardening supplies, trampolines, and basketball hoops.

Overall, American spending on hotels and restaurants plunged by more than one-fifth over the course of 2020. Purchases of airplane, train, and ferry tickets dropped by an even larger margin. Spending on country clubs, golf courses, amusement parks, and other so-called recreational services plunged by nearly one-third. But spending on so-called recreational goods and vehicles—cars, home entertainment equipment, sporting goods—surged by nearly one-fifth. Home furnishings and household equipment—couches, kitchenware, pruning shears—collectively grew by more than 5 percent.

These numbers may not sound large, but in an American economy whose annual output exceeded $23 trillion, even marginal shifts amounted to enormous strain for the factories tasked with making products, and for the ships that carried them across the ocean.

In normal times, an abrupt increase in demand for physical goods was certain to translate into extra orders for Chinese factories. These were not normal times. Much of the rest of the world's industry was still hobbled by quarantine restrictions, which meant that China had become more central to global manufacturing than ever. Even as Chinese factories contended with shortages of labor and difficulty lining up raw materials, they were again open for business. Unprecedented volumes of orders poured into China from around the globe.

The result of this surge was chaos.

Chinese factory managers struggled to find enough people to make all the things that Americans had decided they wanted. Intermittent power outages impeded production as Chinese plants deployed every available production line. Shipping carriers scrambled to redeploy the container vessels they had idled. Monumental traffic jams consumed Chinese ports.

"Everybody wants everything," said Akhil Nair, who oversaw dealings with shipping carriers as an executive at SEKO Logistics in Hong Kong. "The infrastructure can't keep up."

This was the situation confronting Chinese factories as Walker deliberated over the size of the order he should place at Platform88's factory in Ningbo.

Ningbo was a metropolis of more than 9 million people on the East China Sea. Historically, it had been a trading power for much of two millennia, dating back to its days as a major hub on the Silk Road, the network of routes linking merchants in Asia to consumers in Europe and Africa. That status had been revived by the reforms unleashed by Deng Xiaoping in the 1980s. The city became the site of an experiment conducted by Beijing to see what would happen if it streamlined the process of obtaining licenses and approvals to launch new businesses.

This is what happened: foreign investment poured into Ningbo—more than $2 billion a year by 2003, and double that by 2015. The money erected factories that made clothing, auto parts, and home appliances. By the time Walker engaged Platform88, Ningbo was the third largest container port on earth, behind only Shanghai and Singapore.

Platform88's factory in the city had begun life as a state-owned plant that made thermometers. Now, it would focus on *Sesame Street* characters.

Walker acted on Zheng's advice to place the largest order he could manage. In late December 2020, he commissioned 21,196 Elmo figures, and the same number of Julia dolls. The total cost ran to $251,000—the largest purchase in the company's history. In one shot, the Ningbo plant would manufacture 40 percent of the goods that Glo aimed to sell that year.

Walker assumed that his early planning would cushion him against the mayhem roiling the supply chain. The next Christmas season was still a full year away. Surely, that left enough time to get Elmo and Julia to Mississippi.

He did not count on a possibility once difficult to contemplate. The world had all but exhausted the supply of shipping containers.

CHAPTER 6

"AN ENTIRE NEW WAY OF HANDLING FREIGHT"

How a Steel Box Shrunk the Globe

Calvin Zheng was correct. It was indeed a struggle to buy plastic. Migrant workers remained confined to their villages, a reality that continued to stymie Chinese industry.

The electrical outages persisted. And as the country returned to business, the competition for workers and raw materials intensified within China. Fresh orders poured in from Asia and beyond as customers sought alternatives to factories still shuttered in other nations.

Still, the Ningbo factory managed to move ahead with Glo's order. Zheng found a supplier of thermoplastic. He bought LED lights from another factory and even managed to track down basic computer chips.

By May 2021, the Elmo and Julia figurines were nearing completion, leaving half a year to get them across the Pacific in time for the holiday season.

By then, however, Walker was confronting a new and more alarming problem. The price of shipping was skyrocketing. A year earlier,

Walker had been paying less than $2,000 to ship a container load of product from China to the West Coast of the United States. Now, the cheapest option was nearly ten times as much.

And even at those prices, trying to find an available shipping container at a major Chinese port was a maddening exercise in futility. Booking passage had become something like hunting a unicorn.

The shipping container was the workhorse of the global supply chain. It was nothing more than a steel box that could be packed with cargo. But its development in the middle of the twentieth century represented a crucial technological breakthrough.

Shipping containers could be stacked atop others on enormous vessels, then plucked easily off the decks by crane and deposited on the docks. That basic feature had dramatically accelerated the movement of cargo. It had effectively shrunk the oceans, compressing the distance between shores.

There was no global supply chain without the shipping container. There was no Just in Time. The whole system of moving goods around the planet revolved around the container.

All of that had come to pass because of the innovation of a man named Malcom McLean.

McLean was not trying to change the world. He was just looking for a way to avoid traffic.

It was 1953, and the United States was in the midst of a roaring postwar economic expansion. Henry Ford's vision of a car for every household had come to fruition. The result was too many vehicles choking the roads. For McLean, who owned a rapidly growing trucking business, this was both infuriating and expensive.

He had started his venture fresh from high school in the middle of the Depression, using a secondhand truck to move piles of dirt and produce for farmers in the boggy reaches of his native North Carolina. Over the following two decades, he amassed a fleet of more than 1,700 vehicles, making his company the largest trucking business in the South.

His trucks hauled produce from the impoverished farm country of the Carolinas to the affluent consumer markets of New York and Philadelphia. By his midthirties, this enterprise had made him wealthy.

Yet all too often his vehicles were stuck in a traffic jam, or waiting to be weighed, or satisfying some other bureaucratic dictate imposed by a state that sat between his customer and the final destination.

The water beckoned.

The government was sitting on hundreds of cargo ships left over from World War II. It was selling these vessels for pennies on the dollar. Here was the genesis of an idea that would stretch the geographical limitations of the planet.

McLean would buy these surplus war vessels, initially using them to carry truck trailers up the coast to New York City, neatly bypassing the traffic. Then, he improved the plan by ditching the truck trailers, which were not easily stacked, and replacing them with standard-size steel containers. The boxes would soon haul a vast array of cargo.

The key to McLean's innovation was that containers could be filled with freight anywhere—not on the frenzied docks alongside ships, but at factories that made goods, or inside warehouses, or at grain elevators in farming country. Then, they could be hauled seamlessly by truck or rail to a constellation of ports, where cranes specially engineered for the purpose could hoist them onto ships, arranged in tidy stacks.

McLean did not invent container shipping. Others had toyed with the concept for decades. But he was the first to put it to use at significant scale.

Before containerization, the oceans were tempestuous, mysterious, and seemingly limitless. The world's cargo crossed the water in so-called break-bulk ships that had to be loaded and unloaded meticulously, sack by sack, crate by crate, like a three-dimensional jigsaw

puzzle. This process was time-consuming, expensive, and dangerous. It had changed little over the centuries.

The docks were overwhelmed by a riotous assortment of cargo
scattered in ill-fitting configurations—wooden crates, barrels, and
baskets loaded with fruit and meat products alongside steel drums
full of chemical concoctions and piles of lumber plus coils of industrial cables and sheets of steel. Longshoremen navigated this chaos
amid belching trucks, laboring amid the grim assurance that some
would lose limbs as cargo and machinery lurched, crushing anything
in its path. Freight had to be packed and repacked as needed to fit
into truck trailers or railcars for passage to the next stop along the
way, often requiring days.

Each transition—from ship to dock, dock to rail, rail to truck to
warehouse—required a new team of people wielding pluck, muscle,
and courage to fit cargo into whatever receptacle was on hand. Each
stage of shipment exposed the goods to the perils of theft, damage,
and delay.

The container diminished the risks of all of these frustrations. It
reduced the time required to move products from one place to another. And that invited businesses to make their products wherever
conditions seemed most favorable.

The container invited multinational companies to behave as if
the factories of Shanghai and the shopping malls of Houston might
just as well be across the street from one another, with transport
costs and the hazards of the sea substantially reduced.

Big-box retailers and e-commerce giants came to rely on container shipping to stock shelves teeming with an unprecedented
array of choices drawn from factories around the globe.

As most consumers can surmise, the container yielded a bonanza, dropping the prices and increasing the variety of clothing,
shoes, automobiles, and electronics. It put blueberries on supermarket shelves in Chicago in the dead of winter, bridging the distance
from farms in Chile. It placed breweries from Japan to Germany in
reach of beer drinkers from England to Argentina. Teenagers in Los

Angeles donned Premier League soccer shirts, just as people in Manchester sported NBA jerseys.

And the container provided the executive ranks with a potent new tool they could wield to shift the rewards of capitalism away from working people and toward themselves. It reduced the importance of dockworkers, limiting the numbers needed to load and unload cargo. That diminished the power of longshore unions, limiting their ability to threaten trade in pursuit of better pay and job protections.

Beyond the docks, the container gave every multinational company the capacity to credibly threaten to yank factory production somewhere far away if local workers did not bow to their demands.

In a world ruled by the imperatives of Just in Time, the container was the crucial component that made it possible to click a button and expect delivery.

Before the container, globalization had progressed in fits and starts. International trade had expanded dramatically from the middle of the nineteenth century through the early decades of the twentieth, as the seafaring steamship produced its own revolution in cargo. In place of relying on sailing vessels, humanity gained the power to move ships even when the wind wasn't blowing.

Between 1870 and 1910, ocean freight rates plunged by nearly half on American export routes while falling even more dramatically between Southeast Asia and Europe.

In a preview of the impact of the internet a century later, the construction of telegraph wires across the Atlantic supplied merchants from the English port of Liverpool to the American South identical information about the market for commodities like cotton, bringing a convergence of prices and boosting trade.

But the type of globalization propelled by the shipping container was different in magnitude.

By the end of the twentieth century, less than one-third of the containers arriving at the ports of Southern California with imported goods were carrying finished products destined for retailers,

homes, and businesses. Most of the traffic was in so-called interme-
diate goods, items that had been produced in some factory across
the ocean and bound for some other plant that would use them to
make something else.

In short, the container turned the ship into an extension of the
factory.

Longshore workers and seafarers decried the rise of the container
as an affront to the ways of the sea, as the docks were outfitted with
industrial works, and as sailors relinquished their adventures on-
shore to the relentless quest for efficiency. Instead of wandering
New York or Singapore over the days it had once taken for vessels to
be unloaded and loaded again, crews stuck close by the docks. Container ships could be loaded and unloaded within mere hours before
heading back out to the water.

Dockworkers bemoaned the loss of camaraderie and sponta-
neity—along with a dire threat to their pay—as their jobs became
regimented and mechanized, governed by the rhythms of cranes and
forklifts.

"They're turning this job into a factory job," groused one New
York longshoreman.

McLean was unmoved. As he once put it, "I don't have much nostalgia for anything that loses money."

McLean's innovation rested on a key insight about the nitty-gritty
of American economic regulation.

In the middle of the twentieth century, the federal government
was still setting rates for many forms of transportation. Trucking
and shipping were distinct worlds, with widely varying ranges of ac-
ceptable prices.

Railroad operators, fearful of losing cargo business to truckers,
had pressed Congress to keep trucking prices high. That campaign
culminated in the passage of the Motor Carrier Act in 1930, which
limited how many trucking businesses could enter the marketplace.

Shipping, on the other hand, was treated as an endangered industry, one that had been losing freight business for decades. The government allowed shipping companies to charge much lower rates. McLean's idea for putting trailers onto ships was a savvy ploy to secure an advantage over competing trucking firms. He would capture their business by undercutting them on price, using ships to gain the right to charge as little as sea cargo companies.

The more boxes he could pack on a single vessel, the lower his costs of moving cargo per ton.

One of McLean's customers was a prominent brewer, Ballantine Beer. Transporting a load of beer from the New York area to Miami on a cargo ship then cost about $4 a ton. Putting the beer into a trailer at the brewery and hoisting it onto one of McLean's new vessels could get the same load down to Miami for a mere 25 cents a ton—a savings of more than 90 percent.

But where could McLean find the boxes?

His team located a manufacturer in Spokane, Washington. It produced thirty-foot aluminum containers deployed on barges running between Seattle and Alaska. He ordered a pair and had them delivered to the Baltimore shipyard that was reconfiguring his tankers. He and a handful of other executives subjected the boxes to a rigorous, scientific testing regimen. They climbed atop the containers and jumped up and down. Satisfied that their roofs could withstand the pounding, McLean soon ordered two hundred.

The hub of McLean's fledgling venture was located in Newark, a backwater port in New Jersey, across the harbor from the teeming docks of Manhattan. There, he found an accommodating government arm in the Port Authority of New York and New Jersey, an agency forged by those two states to promote regional trade. The port authority was eager to elevate Newark. McLean quickly recognized its value as an alternative to New York City.

The docks of New York were notoriously crowded, decrepit, and choked by traffic. They had long served as the gateway to Europe for local factories, whose operations were strategically clustered nearby.

But as manufacturers in the Midwest expanded their sights globally, New York's docks were rapidly losing appeal as a jumping-off point across the Atlantic. The rail lines that brought in goods from the rest of the country terminated in New Jersey. A batch of machinery arriving from a factory in Indiana and bound for Europe by ship had to be off-loaded at a rail yard and then carried by barge to the piers of Manhattan. Freight that came in by truck had to enter the city via a tunnel and then navigate congestion on the narrow streets leading to the docks.

Newark, by contrast, had abundant room for expansion, immediate proximity to the rail yards, and direct access to a shiny new highway called the New Jersey Turnpike. And the port authority had the power to sell bonds and build the shipping terminal that McLean required. He could lease the facilities while saving his capital to build out his fleet of ships and containers.

On April 26, 1956, local officials gathered at the Port of Newark to gawk at the spectacle of a crane depositing containers—one every seven minutes—onto the first of McLean's retrofitted tankers, the 524-foot *Ideal-X*. Less than eight hours later, the vessel, loaded with fifty-eight boxes, departed on its inaugural run to Houston.

The chairman of the port authority was in attendance. So was the mayor of Newark, celebrating a great leap in economic development.

But one key observer looked out at the *Ideal-X* with contempt.

"I'd like to sink that sonofabitch," declared Freddy Fields, a top official at the International Longshoreman's Association, the union that represented East Coast dockworkers.

Fields was right to divine a threat to his union brethren. Before the container, roughly half the costs of moving freight by sea were paid out in the form of wages to dockworkers. McLean's new system represented a direct assault on that equation. On the day the *Ideal-X* first set sail for Houston, loading a cargo vessel by hand cost nearly $6 a ton. Containerization soon dropped the cost to 16 cents per ton.

Through the 1960s, McLean added new, higher-capacity cranes to his terminals while dreaming up novel ways to squeeze more boxes onto the decks of his ships. Like Henry Ford and Taiichi Ohno, he was intent on excising waste from the process, speeding up the loading of ships and the transfer of containers onto truck beds and rail.

"McLean's fundamental insight, commonplace today but quite radical in the 1950s, was that the shipping industry's business was moving cargo, not sailing ships," wrote the economic historian Marc Levinson in his indispensable book about the shipping container, *The Box*. "McLean understood that reducing the cost of shipping goods required not just a metal box but an entire new way of handling freight. Every part of the system—ports, ships, cranes, storage facilities, trucks, trains, and the operations of the shippers themselves—would have to change."

By 1961, McLean's company—by then renamed SeaLand Service— was in possession of four surplus World War II tankers. Each had been extended to carry 476 containers, an eightfold increase over the *Ideal-X*.

The following year, SeaLand opened a terminal at a soon-to-be container colossus at Port Elizabeth, on marshland just south of Newark. Soon, McLean secured the government's blessing to run containers from Newark to California via the Panama Canal.

Through the end of the decade, competing carriers launched container services to Europe, and then to Asia and Latin America, as break-bulk shipping began to disappear and international trade swelled.

But a major driver of containerization had little to do with commerce. The American war in Vietnam demanded a more efficient way to move cargo across the ocean.

By June 1965, the United States had deployed nearly 60,000 troops to Vietnam. The Pentagon was planning a dramatic buildup in forces, and requested more than 400,000 additional troops by the end of that year. Supplying them with weaponry, ammunition, and provisions was a colossal challenge.

Vietnam's lone deepwater port, on the Saigon River, could not handle the dramatic influx of cargo arriving from California. Its ten berths were overwhelmed. Its docks lacked cranes.

Vietnam's other harbors were notoriously shallow, making it impossible for ocean craft to pull up to docks for unloading. Incoming vessels dropped anchor offshore, sometimes miles from land. Their crews had to lower cargo piece by piece into barges that would float them to shore. Weeks transpired before the entire contents of some vessels could be ferried to land.

Delegations of American officials visited Vietnam hoping to celebrate battlefield victories, only to hear ominous warnings from local officers that the war effort was being impeded by logistical snafus. McLean persuaded them to let him take a crack at a fix.

Just before Christmas in 1965, he flew into Saigon and toured the country's largest ports. True to form, he recommended that the solution was containerization.

The reward for his counsel was a $70 million contract for SeaLand to take over key pieces of the American shipping effort. He deployed six ships to carry cargo from Oakland and Seattle to Vietnam.

McLean's undertakings soon unclogged the supply chain. That success did not spare American forces from a tragic fiasco in Vietnam. But it did advance the spread of container shipping across the Pacific, especially as SeaLand divined a way to bolster its profits on the run to Vietnam. It would stop in Japan on the return leg.

Japan was sending growing volumes of industrial equipment, electronics, and automobiles to the United States, making it an enticing market for container shipping. The Japanese government was already subsidizing the development of container terminals in Yokohama and Kobe.

By 1968, SeaLand was running six ships per month between Japan and the West Coast. This business amounted to free cash dispensed by Uncle Sam. The military was already paying enough on the outbound leg to Vietnam to cover the whole journey.

Soon, the Pacific would be crowded with scores of container vessels run by a growing assortment of carriers.

Between 1966 and 1983, at least 122 countries joined the emerging international container shipping network, loading their first boxes at ports or rail yards. Among the wealthiest industrial economies, container shipping expanded trade nearly ninefold over the first two decades.

The ports at the center of this refashioning were increasingly winner-take-all affairs. As container ships grew ever larger and more expensive to build—eventually exceeding $100 million per vessel—carriers sought to maximize their time on the water by limiting their ports of call. They stopped only at the largest docks with the most advanced systems for loading and unloading, and the easiest access to highways and rail. The others receded into history.

Newark rose to become the most important center of shipping on the American East Coast while the docks of Manhattan were left to disintegrate. On the West Coast, the port of Oakland embraced containers and turned the docks of San Francisco into relics.

In Britain, the otherwise unremarkable town of Felixstowe decimated the docks of London, turning itself into the nation's largest container port. This was a triumph for the consultants at McKinsey, who counseled the British government to concentrate much of the nation's container shipping in one enormous port—the principles of *lean* unleashed on the water.

Across the English Channel, Rotterdam recovered from the devastation of World War II by constructing Europe's largest port, with container shipping at the center of its designs.

Taiwan and Hong Kong invested aggressively to develop container ports. Singapore adopted containerization and became the dominant so-called transshipment point serving Southeast Asia. Cargo arrived from around the globe and was transferred onto smaller vessels and dispatched to Thailand, Indonesia, and elsewhere.

By the middle of the 1980s, the largest container vessels had room for 4,000 twenty-foot containers—nearly seventy times as many as on the *Ideal-X*. Fifteen years later, the biggest ships were moving as many as 8,000 boxes. Two decades after that, ships capable of carrying more than 23,000 containers were plying the seas.

By then, a world that reveled in records was absorbing the spectacle of enormous container ships with near-pornographic fascination. This ship was larger than four football fields; that one used more steel than the Chrysler Building.

No nation invested more aggressively in container shipping than China. Two decades after entering the WTO, the state-owned China Ocean Shipping Company—better known as Cosco—was the fourth largest container carrier on earth. Chinese companies made 80 percent of the cranes used to load and unload the boxes. Seven of the world's ten busiest container ports were in China, including the largest of all—Shanghai.

This was the context that had allowed a company in a small town in Mississippi to depend on factories on the other side of the Pacific to make its goods.

Hagan Walker had previously shipped his orders in quantities small enough to fill only part of a container. Now, for the first time, he was purchasing enough products to fill his own forty-foot box.

There were roughly 50 million shipping containers in operation worldwide. Walker needed precisely one of them. Yet even that was suddenly an impossible ask.

CHAPTER 7

"CARRIERS ARE ROBBING SHIPPERS."

The Floating Cartel

By the time Walker began trying to book a container to carry his *Sesame Street* order across the Pacific in May 2021, shipping prices had soared. Some of Platform88's other customers were stashing their finished goods in warehouses in China, waiting and hoping that prices would come down.

For Walker, waiting was not an option. He needed to secure space on a container ship immediately or he risked missing the holiday season.

He emailed a shipping agent he had dealt with previously, a China-based company called ECU Worldwide. They offered to move only a partial container—one-third of a forty-foot box—from Ningbo to Starkville for $5,485. That would mean leaving behind most of his order for later, which was far from ideal. But it was better than nothing. He quickly agreed.

Four days later, the same agent emailed Walker to report that she could not confirm the booking. She simply could not locate an empty container anywhere near Ningbo.

Walker began firing off emails en masse.

He heard back from Jefferson Clay, a shipping agent at an outfit called Cargo Services in Indianapolis, who urged him to forget about moving his freight out of Ningbo. He could instead truck his goods ninety miles up the coast to Shanghai, where boxes would be easier to find.

"Ningbo has zero space lately," Clay wrote in July. "Truthfully, it's all a mess, and the pricing is out of this world."

Two days later, Clay emailed again. Nearly every conceivable route to Mississippi was a no-go. Carriers were balking at allowing one of their containers to be sent inland, hundreds of miles from a port. With cargo prices still exploding, they were eager to return empty boxes back to China as quickly as possible to grab the next load of factory goods.

Clay proposed that Walker consider booking a ship to Houston, or perhaps Mobile, Alabama, on the Gulf of Mexico, and then truck the goods to Starkville. But he offered no assurances that he could find a vessel.

He also suggested that Walker consider breaking his order into four separate shipments, and then book each as a partial container, even as he added that this could multiply his problems.

"The delays and congestion on those four could be disastrous," he wrote.

Faced with this assessment, Walker contacted still another agent, Harry Wang at Baylink Shipping, a company based in New York. He suggested that Walker try his luck at the port in Shenzhen, which would entail trucking his goods nearly nine hundred miles from the factory in Ningbo.

But he had to move fast, because prices were still rising.

"China now is crazy," Wang wrote. "Carriers are robbing shippers and importers."

The most immediate explanation for the turmoil in the shipping industry was obvious: a crippling shortage of shipping containers.

As Americans filled their homes with goods to endure the pandemic, Chinese factories were pumping out such enormous volumes that they were exceeding the supply of boxes at nearly every port.

The largest retailers like Amazon and Walmart had contracts with ocean carriers that essentially guaranteed their ability to place their containers on board vessels. And when the situation got so dire that even these giants struggled to get cargo on ships, they were able to charter their own container vessels and keep their products moving.

But for smaller players, the container shortage posed an enormous threat to their bottom lines.

Baum-Essex, a brand unknown to the consumer masses, was a force in putting familiar products on shelves. Headquartered in New York's Empire State Building, the family-run company relied on a network of factories in China and Southeast Asia to manufacture umbrellas for Costco, cotton bags for Walmart, and ceramics for Bed Bath & Beyond. As recently as the summer of 2020, the company had been paying as little as $2,500 to ship a forty-foot container from Asia to the West Coast of the United States. Six months later, the price had more than doubled, while continuing to climb.

"I just signed off on a bunch of containers to California and we just paid $6,700," the company's chief operating officer, Peter Baum, told my colleague Alexandra Stevenson in early 2021. "This is the highest freight rate that I have seen in 45 years in the business."

Many of the products Baum was making in China were already finished and ready to sail, but the ocean carriers were telling him that they had no space on their ships. His goods were sitting on the factory floor waiting. One container full of wicker chairs and tables had been stuck on the dock of a port for ninety days before eventually getting loaded onto a ship.

As far away as Ireland, companies that exported computer parts, pharmaceuticals, and medical devices were paying $9,000 to send a forty-foot container to China—more than four times what it had cost only three months earlier.

"We have a dramatic situation that hasn't been seen for decades in terms of this acute shortage of containers," John Whelan, an international trade consultant in Dublin, told me.

In reality, I discovered as I reached out to shippers, containers were not in short supply so much as scattered in the wrong places. Empty boxes were piled in towering stacks near container terminals in Australia and New Zealand. Yet containers were so scarce at India's port of Kolkata that local factories were trucking electronic parts more than one thousand miles west, to the port of Mumbai, where they could be found in greater numbers.

In Thailand, Vietnam, and Cambodia, containers were so difficult to locate that rice exporters were scrapping shipments to the United States. Yet in Los Angeles and Newark, containers were so plentiful—and still piling up—that port operators were scrambling to develop emergency storage yards.

Part of the trouble stemmed from China's dominance in making medical protective gear like face masks and gowns. During the first months of the pandemic, container ships carried staggering volumes of these products to points around the globe. In regions like West Africa and South America, where vessels landed less frequently, containers frequently sat uncollected after they were unloaded. As carriers eliminated stops and reduced capacity during the first wave of the pandemic, these wayward boxes remained stuck. Over the first two months of 2020, major carriers scrapped roughly one-fifth of their ports of call in Sub-Saharan Africa and a slightly smaller share of stops in Latin America.

And even as carriers retrieved idled ships later that year, they concentrated their fleets on their most profitable routes linking Asia to Europe and North America. Elsewhere, stacks of empty boxes sat awaiting ships that could carry them to where they were needed.

The effect of this was to reduce the supply of containers at ports in China. An empty box stuck at the bottom of a stack in South America or West Africa was a box that could not be loaded in Shanghai, Ningbo, or Shenzhen.

Efforts to increase the number of containers collided with the same problem that challenged the rest of the global supply chain. Three Chinese companies dominated production, manufacturing 96 percent of the world's ordinary shipping containers, and all of the refrigerated variety.

The largest, China International Marine Containers, was a state-owned enterprise, meaning that it answered directly to the strategic imperatives of the Chinese Communist Party. The same company effectively controlled the China Container Industry Association, a trade group that influenced the operations of the rest of the manufacturers. The second largest, Dong Fang International Containers, was controlled by Cosco, the state-owned shipping giant.

The factories that made shipping containers were reportedly ramping up production, but they were limited by the same challenges impeding the rest of Chinese industry—not enough workers, not enough electricity, rising costs for raw materials.

They were also guided by a new consideration: maximizing profits.

For years, the three Chinese manufacturers had pumped out as many containers as they could, largely disregarding prices as they served the broader ambitions guiding national economic policy. More containers meant more capacity to deliver Chinese exports to world markets.

But in more recent times, the container manufacturers had focused on enhancing their balance sheets. That limited the pace of their production. Making fewer containers kept them scarce enough to ensure that buyers would have to compete with one another to amass a supply, elevating prices.

"The factories are behaving differently than they have in the past," said Tim Page, who ran a company that leased shipping containers to carriers. "They don't have any interest in increasing production at the expense of price."

A container shipping analyst at an industry data firm, Drewry, described an organized form of collusion.

"The three main companies decided to try to support some kind of pricing for the dry freight boxes," said the analyst, John Fossey. "They basically got together and said, 'Look, we're not going to produce these containers at a loss.'"

Here was a feedback loop of supply chain duress. The advent of the shipping container and the corporate quest for lower costs of production had combined to make the global economy dependent on Chinese factories for practically everything. And as the system broke down, the only immediate fix was greater reliance on China.

At the same time that containers were scarce in key places, major ports around the globe were contending with the ravages of the pandemic itself. Dockworkers from Shenzhen to Seattle were stricken by COVID-19 and forced to quarantine, limiting available hands. From Newark to Ningbo, truck drivers were falling ill, leaving loads uncollected. Warehouses near every port were jammed with cargo that could not be moved in or out because of a shortage of available labor.

"I've never seen anything like this," declared Lars Mikael Jensen, head of Global Ocean Network at A. P. Moller-Maersk, the Danish shipping goliath. "All the links in the supply chain are stretched."

Such were the strains already confronting the supply chain when Americans began drawing on it more aggressively than ever, turning their homes into pandemic sanctuaries.

Between September and November 2020, the volume of exercise equipment moved by container from Asia to North America more than doubled compared to the previous year. Shipments of cooking equipment almost doubled over the same period, while containers full of disinfectants grew by 6,800 percent.

By June 2021, one of China's largest container ports—Yantian, near Shenzhen—was partially shut down by a fresh outbreak of COVID-19, adding to the global disruption by waylaying more boxes.

The number of containers had effectively been diminished, just as the need for the boxes was growing. The basic laws of economics dictated that the price of shipping was going to rise.

"It's a classic supply-and-demand issue," said Kim Bradley, whose Massachusetts-based company, Highline United, imported shoes from factories in China for brands like Isaac Mizrahi.

But there was a broader context to all this, another reason why ocean freight prices were multiplying without limit.

The shipping business was opaque, lightly regulated, and dominated by a handful of international companies that held a monopolistic grip over the marketplace. They were perfectly positioned to exploit any disruption on the seas.

In an age in which people around the globe depended on container ships to summon most of their goods, the industry that controlled the vessels functioned much like a cartel.

The successful profiteering by the ocean carriers of a global calamity stemmed from their triumph in having extracted themselves from government oversight.

In the United States, the ocean cargo industry had for decades been regulated like a public utility, with transparent prices that were, by law, available to every potential shipper. This state of play flowed from the Shipping Act of 1916, which created a federal body, the United States Shipping Board, to ensure fair competition.

Under the act, ocean carriers were exempted from American antitrust law. This permitted them to form so-called conferences—basically, alliances that forged shared routes and prices, governed by the idea that this was the best way to ensure dependable service. In exchange for the right to coordinate their schedules, the carriers had to submit to stringent policing of their prices.

The act compelled carriers to disclose their contracted arrangements to the shipping board, which had the authority to block deals that it construed as threats to fair competition. The law gave the board power to "disapprove, cancel, or modify any agreement" that it determined was "unjustly discriminatory or unfair as between carriers, shippers, exporters, or ports." And it specifically barred carriers from denying passage on their vessels when space was available.

In practical terms, the shipping act obligated carriers to treat every shipper equally. They were forbidden from negotiating sweetheart prices with particularly large customers—offering rebates and guaranteed access to their fleets—while denying those terms from some other shipper. Every business that needed to move goods across the water was entitled to the same deal, much as every household in a city could count on paying the standard, government-regulated rate for electricity or water.

But that mode of operation—heavy on bureaucratic interference—could not withstand the deregulatory revolution that would seize Washington.

By the early 1980s, with Ronald Reagan in the White House, American corporate interests had succeeded in eradicating a vast range of federal strictures. Lobbyists used the cold war and the specter of Soviet-style Communism to sell deregulation as a distinctly American version of freedom. They depicted price controls, antitrust enforcement, and prohibitions on industry collusion as the trappings of state-dominated authoritarianism.

This was the spirit in which Congress deregulated much of American transportation. Under the Shipping Act of 1984, the antitrust exemption that had long allowed carriers to set prices and routes was extended to other players engaged in moving shipping containers, including the railroads and trucking firms.

By then, a new regulator, the Federal Maritime Commission, had oversight of shipping, but the 1984 law significantly limited its authority. The routes and pricing agreements forged by carriers no longer required federal approval. The only way the commission could block a deal was to file a lawsuit and then persuade a court that competition was clearly being harmed.

One key vestige of the old system remained. The conferences still had to publicly disclose their prices and routes, maintaining some semblance of transparency. But that component was eliminated the following decade, with the passage of the Ocean Shipping Reform Act of 1998. The ocean carriers retained their antitrust exemption, extending their right to coordinate their routes. They also gained the power to negotiate deals in secret, offering better prices and priority service to their largest customers.

Some warned that the change would condemn the supply chain to domination by the largest companies at the expense of fair competition.

"The bill's provisions allowing completely secret contracts go too far, and risk discrimination and abuse adverse to US trade interests," declared the then chairman of the Federal Maritime Commission, Harold J. Creel Jr., during a 1997 congressional hearing.

But discrimination was the very point of the alteration. The largest American importers lobbied for the change, cognizant that the innate advantages of scale were being held back by a level playing field.

Under the old system, companies like Walmart and Home Depot—which were bringing thousands of containers a month across the water from Asia—paid the same rates as small businesses that moved only a few boxes a year. Once the market was deregulated, the behemoths could use their size to extract discounts, while securing guaranteed access.

The ocean carriers exploited their new freedom from regulation to undertake a merger binge. The largest fleets gobbled up smaller competitors in a perpetual reach for greater scale. McLean's old company, SeaLand, was taken over by Maersk, the Danish shipping behemoth, in 1999. Amid a vicious price war, many carriers succumbed to bankruptcy. The survivors divided themselves into three alliances—like those created by major airlines—pooling their bookings and sharing freight.

By 2018, the three alliances controlled 80 percent of the global container shipping market, up from about 30 percent a decade earlier. On the most lucrative routes—those crossing the Pacific—they commanded 95 percent.

For the first two decades following the 1998 deregulation, the growing market power of the largest carriers was rendered insignificant by the motives of the nations that subsidized them. From mainland China to South Korea to Taiwan, governments lavished credit on carriers, giving them the finance to build out their fleets. This also spurred domestic shipbuilding along with attendant enterprises like steelmaking. They aimed to keep shipping prices low to stimulate exports. The bottom-line concerns of the carriers themselves were set aside as peripheral to the broader interests of national development.

For multinational retailers, this was like fertilizer for their ambitions. They relied on the easy availability of container vessels to construct Just in Time supply chains spanning oceans, while treating transportation as a trifling cost. From Europe to the United States, consumers reaped the gains via inexpensive imports. Zara, H&M, Uniqlo: all benefited from monumental scale and cheap shipping rates.

But somewhere between Malcom McLean's *Ideal-X* and another now-infamous ship, the *Ever Given,* the merits of scale began to work in reverse.

The *Ever Given* was loaded with eighteen thousand containers when it hit the side of the Suez Canal in March 2021 and remained lodged there for six days.

The episode captivated the world, providing an almost comical encapsulation of how the global economy had gone off the rails. People gawked at televised footage of the massive ship lying alongside a concrete embankment, motionless. On social media, the *Ever Given* became the inspiration for thousands of memes and video clips— images of a giant excavator working the mud on the edge of the canal, now dwarfed to insignificance by the adjacent hull of the ship; cartoons of hot air balloons used to lift the vessel clear. In a world confused by the disappearance of everything from medical gear to gadgets, here was a window onto what had happened. Here was a powerful illustration of the fragility of the global supply chain. The breakdown of this single ship was halting $10 billion a day of seaborne trade.

Given that nearly one-third of the world's container cargo passes through the Suez Canal—more than $1 trillion worth per year—the consequences of the shutdown swiftly rippled out to ports from Rotterdam to Newark to Shanghai. Every ship that arrived late to one port as the result of the Suez closure was a vessel that could not be loaded with someone else's cargo. Delays triggered more delays. Chemicals stuck in containers in the center of Egypt impeded the production of paint at factories in Pennsylvania. Loads of auto parts not delivered on time to Germany spelled cars that could not be made.

Yet to those who had been paying attention, the incident was less a shock than a confirmation of warnings that experts had been issuing for years.

Back in 2015, the International Transport Forum—an intergovernmental think tank that included fifty-four members—sounded the alarm about the deteriorating benefits of so-called megaships. Over the previous decade, the increased size of container vessels had reduced the costs of moving a box by about one-third. But those savings were disappearing as ever larger vessels demanded additional dredging of harbors, larger docks, bigger cranes, and the raising of bridges—all substantial costs.

Megaships were limiting choices for shippers, forcing them to rely on carriers that possessed extreme market power given their inclusion in one of the dominant alliances. Enormous ships were concentrating freight at a smaller number of ports—those with the capacity to handle the largest vessels.

Containerization had been a major driver of Just in Time manufacturing, allowing factories to schedule deliveries of parts as needed. But unbridled scale was destroying the efficiencies. As ships grew ever larger, ports could no longer count on a regular flow of cargo to unload. Now, they were confronted by thousands of boxes jamming the docks all at once on the days that a huge vessel pulled in—necessitating more longshore workers, more truck drivers, and more railcars. On other days, the ports were quiet.

All of this left the world economy vulnerable to chaos whenever some unforeseen shock arrived. With the container fleet controlled by a handful of carriers whose megaships were confined to the largest ports, they were positioned to dramatically increase their prices in event of any disruption. Shippers would have no alternative but to pay.

It would take the pandemic to bring home the truth of that warning.

Columbia Sportswear, the outdoor clothing giant, was an archetype of globalization. Based just outside Portland, Oregon, the company maintained a down-to-earth demeanor even as it grew into a major international brand. Its designers were clustered in the Pacific Northwest. Its factories were in Asia. Container ships bridged the divide. Its executives had long operated as if inexpensive and reliable ocean cargo was an immutable part of the world economy.

"It's been something that the company historically hasn't really worried about," Columbia's chief executive officer, Timothy Boyle, told me. "It's sort of like every day, when you get up in the morning, you turn on the lights, and the lights always work. The logistics infrastructure was always something that was cheap and available."

But as we spoke in August 2021, Columbia was having to pay as much as $25,000 to move a container of goods across the Pacific—a tenfold increase compared to a year earlier. And the company was frequently being told that there was no space on board for its shipments.

Boyle was looking ahead to the holiday season and concluding that delays and product shortages were inevitable. He was reconsidering the merits of depending on faraway factories.

"It's a question of how long this lasts," he said.

The following year, American importers were still bemoaning

their inability to get freight onto ships, even when they had contracts with carriers that seemingly secured such rights.

Consumer prices were soaring on a vast range of goods, from clothing to furniture to groceries.

And the largest container shipping carriers were on track to log $300 billion in profits for the year, on top of more than $200 billion the year before.

None of these things appeared coincidental.

"It's just them manipulating the market to see how they can drive the price," said Jason Delves, who ran a Tennessee-based company that imported flooring, cabinetry, and outdoor furniture, most of it from Asia. "Contracts are not worth the paper they are written on these days."

Delves's company, F9 Brands, typically moved fifty containers a week full of cabinets and flooring materials from China, Malaysia, Vietnam, Indonesia, and Thailand, along with boxes full of carpets from Dubai. He brought his cargo into a container port at Savannah, Georgia, in order to bypass the legendary traffic jams at the ports on the West Coast.

Some of his containers traveled west through the Suez Canal. Most moved east, across the Pacific from China, eventually passing through the Panama Canal, and then up the Atlantic coast.

But by the fall of 2021, just as Walker's order was floating off Long Beach, his boxes were not moving at all. The carriers were refusing to put them on their ships.

His two furniture brands—Cabinets to Go and Gracious Homes—had agreements with carriers that were supposed to protect him from this eventuality. The contracts locked in an agreed-upon price—an average of $6,970 each—to move 1,040 containers from China and Southeast Asia to Savannah between May 2021 and April 2022.

But over the course of that year, the two brands managed to get only 166 boxes—about 15 percent of their allocation—on board ships at the contracted rate.

The rest of the time, the carriers said there was no space on board, or no container could be found, or some other justification for leaving Delves's boxes stuck at a dock or in a warehouse somewhere in Asia, full of furniture that was supposed to be in the United States.

That forced Delves to venture into the so-called spot market, paying whatever the going rate was at that moment, to get his containers onto ships. His company had managed to transport 355 boxes in this fashion, paying an average of $15,350 each—more than double its contracted rate.

Frequently, the same carriers that said they had no space on their ships suddenly found room—often on the very same vessels—provided Delves was willing to pay for "premium service" or "super-premium," or some other invented term that meant he was going to hand over gobs of additional money. He had moved 163 boxes in this way, paying an average price of $22,500, more than triple his contracted rate.

"The only thing that premium and super-premium guarantee is that you are paying more for that container," Delves told me. "If we were doing what they are doing, we'd get arrested."

To the extent the carriers responded to such accusations, which was almost never, their people would tell you—never on the record—that the contracts were loaded with caveats that made them unenforceable. This had once been a benefit to those now complaining.

Back in the days when there was too much capacity, importers exploited the flexibility of contracts. Their deals obligated carriers to move a minimum number of boxes at a set price. But if the customer opted to move fewer, they did not have to pay a penalty.

Now, the dynamic had reversed. Supply was tight, prices were astronomical, and the carriers were behaving like miners unleashed on a gold rush. The niceties of their previous dealings were ditched in the pursuit of a frenzied reach for lucre.

"This is arguably the largest driver of the increased cost of consumer goods in our country," Delves said. "This surpasses any tariff that's put on anything."

There were certainly other factors behind soaring prices. Governments in major economies had dispensed cash to their citizens to help them manage the economic strains of the pandemic, which had boosted spending power. Decades of consolidation in many industries—from meatpacking to telecommunications—had placed companies in position to exploit disruptions as an opportunity to lift prices.

But the shipping industry's windfall gains were clearly a significant driver. Increased ocean freight rates in 2021 lifted the price of goods worldwide by some 1.5 percent into the following year, the International Monetary Fund estimated.

In Chicago, David Reich reluctantly concluded that his import business was too small to matter to the ocean carriers who monopolized the routes from Asia.

His company, MSRF, assembled gift baskets full of coffee and hot cocoa for retailers like Walmart and Walgreens, importing key elements such as bowls and mugs from China. But as the holidays approached in 2021, he found that he could not get his boxes on board ships even after agreeing to pay premium charges.

A South Korean carrier, HMM—formerly known as Hyundai Merchant Marine—delivered only nine of the twenty-five containers he had been promised in his contract. Yang Ming Marine Transport, a Taiwanese firm, moved only four of the one hundred boxes he thought he had secured in a contract with that carrier.

"We are finding it impossible," Reich told me. "It is just brutal."

His contracts were then up for renewal. Reich was incensed to learn that the companies were uninterested in even discussing new deals. They were focused on serving the needs of their largest customers—Amazon, Walmart, and the other enormous retail operations that seemed set to emerge from the mayhem in even stronger position.

"They said, 'Sorry, we're too busy,'" Reich told me. "They have taken care of the big customers, and there is no room at the inn."

Within the shipping industry, it was by then an open secret that the world's largest retailers were among the beneficiaries of the chaos roiling the seas. They alone could afford to charter their own vessels, accepting extra costs as a means of capturing sales from competitors whose boxes were stranded.

This was reflected in a revealing interview that appeared in the trade journal *American Shipper,* conducted by journalist Greg Miller with Lars Jensen, a former Maersk executive who headed a consultancy called Vespucci Maritime (and a different Lars Jensen from the Maersk executive quoted earlier).

"You now have a massive competitive advantage compared to your smaller competitors," Jensen said. "If I were a large importer I would not be complaining about this situation. Sure, I would be somewhat frustrated that I have to pay three or four times more than I did last year, but my competitors are paying ten times more.

"I would look at this as a strategic opportunity," Jensen continued. "I would eat that loss myself and not increase my retail prices, because I can afford to absorb this, and my competition cannot. I would basically drive my competitors out of business."

That the carriers were able to command exponentially higher rates for shipping was itself proof of the merits of taking a far-sighted approach. They had lost money for years while expanding their market share. They had eliminated competitors through mergers. This had placed them in the ideal position to lift prices once something happened to make the market tight.

"We have definitely seen the effects of consolidation," Jensen said. "There is a de facto oligopoly."

Jensen was giving away the game. The monopolistic predations of the container industry had presented the world's largest retailers with an enticing opportunity. They could ride out the storm while watching their competitors drown. They would survive to enter a new era in which they would have an even greater hold over their own markets.

In the global supply chain, everything was intertwined. A monopoly in one place created the conditions for a monopoly to develop somewhere else. Which meant that smaller players had an increasingly large problem.

This was why Hagan Walker was now staring at a full-blown crisis.

By June 2021, Baylink, the shipping agent in New York, was telling Walker that, in theory, he could move his container from Shenzhen to Houston for $21,500. Except there was no guarantee that the journey would ever be confirmed.

"Honestly," wrote the agent, Harry Wang, "overseas agents in Shenzhen can barely make bookings for space as most of the ports in China are suffering from terrible congestion."

There was also the not-incidental matter of bridging the final 569 miles separating Houston from Starkville.

"Truckers in Houston are fully booked at this moment," Wang wrote. "We are not sure if they will take such a long haul."

Cargo Services, the outfit in Indianapolis, offered to book a forty-foot container to Mobile, Alabama, from the port of Yantian for $22,532. In addition to trucking the cargo there from Ningbo, Walker would have to wait an indeterminate period for a vessel that had space.

"Honestly," the import manager wrote, "we are seeing the soonest available vessel space from China mid-September."

Walker used the Chinese social media platform WeChat to contact a freight handler in China, Seabay.

There, a representative named Sunny Liu warned him that the market was still getting tighter. She urged him to book whatever vessel he could on any route reaching any American port, while leaving for later the question of how to complete the journey to Mississippi.

On August 30—more than three months after his search began—Seabay provided a confirmed booking: Shenzhen to Long Beach for $28,296, with an estimated delivery date of October 30.

Normally, such quotes were good for thirty days. This one expired in twenty-four hours, an indication that prices were still rising fast.

Walker paid immediately. Seabay arranged to have a container delivered to the Ningbo factory.

By then, a new source of anxiety was tormenting Walker. On the other side of the ocean, at the ports in Southern California, chaos was building to alarming levels, threatening more delays even after his shipment managed to complete the crossing.

Dockworkers were succumbing to COVID-19. Truck drivers were in short supply. The American rail system was buckling. Containers were piling up on the docks. Scores of vessels were stuck floating miles off the coast, waiting for an open berth.

Walker imagined his container getting jammed there. He looked for a means to eliminate sticking points.

Given the difficulty in securing containers, Walker had opted to cram his box full of his products from top to bottom—a practice known as floor-loading—rather than placing them on top of shipping pallets.

Pallets were handy. You could stick a forklift through them and move a whole stack of products around a warehouse. But pallets also took up precious room inside containers. Making maximum use of space now overrode all other considerations. So Walker had dispensed with pallets, while filling every inch of his container with his *Sesame Street* order.

Yet the more he heard about the madness awaiting him in Southern California, the more he regretted that decision. Without pallets, his container would need to be unloaded by hand at a warehouse near the port, a process that would require several workers and consume days.

He contacted Sunny Liu at Seabay. Could he rework his order to add pallets?

Too late, she told him.

Then she texted him about another dismaying development. Chinese authorities had shut down part of the port of Ningbo after a single dockworker tested positive for COVID-19. Cargo was being diverted en masse to Shenzhen, making it even more difficult to find a place to store containers awaiting vessels.

Walker had to get his box to Shenzhen within three days. Otherwise, his space on the ship would be given to someone else.

Walker relayed this news to Platform88. He was told that it would be extraordinarily difficult to find a truck driver to haul his container to Shenzhen on short notice. Many had containers stuck behind their rigs that they were unable disgorge given the closure of the Ningbo port. That meant they could not pick up a fresh load.

Somehow, Zheng's team located a driver. Glo's container was finally on the move, headed toward Shenzhen on a seventeen-hour highway excursion.

The box was loaded atop the *Maersk Emden,* one of more than three hundred such vessels owned by A. P. Moeller-Maersk, the Danish shipping conglomerate.

In an industry defined by scale, Maersk was a behemoth.

By itself, Maersk moved about 17 percent of all shipping containers worldwide. And it was part of an alliance with the world's largest container carrier, Mediterranean Shipping Company, which alone operated more than seven hundred vessels.

The *Emden* glided out of Shenzhen on September 12.

The vessel stopped at Nansha, near the Chinese city of Guangzhou, and then at Yantian, east of Shenzhen. She docked at the port of Ningbo, by now reopened.

On September 27, she embarked across the unfathomable expanse of the Pacific.

PART II

Across the Water

CHAPTER 8

"THE LAND OF THE FORGOTTEN"

How Farmers Got Stuck on the Wrong Side of the Water

Hagan Walker had it better than the most aggrieved shippers. Even as he struggled to line up a container in China, and even as he agreed to astronomical prices, he was at least operating in a marketplace that was the highest priority for the ocean carriers. Their routes from China to the West Coast of the United States were their most heavily trafficked and their most lucrative. The carriers were placing as many containers there as they could muster.

Meanwhile, shippers almost everywhere else were prone to being ignored, their cargo stranded, their sales plummeting.

On the opposite edge of the Pacific, in the Central Valley of California, Scott Phippen's orchards were bursting with almonds. Ordinarily, this was a satisfying sight. Here was money growing on trees.

But in the spring of 2022, the usual spectacle of branches erupting with nuts came tinged with a sense of foreboding.

Phippen headed a clan that was among the most important almond growers in California, a state that produced 80 percent of the world's supply. As he contemplated a harvest that lay only a few months ahead, he could not shake the fear that he might run out of places to stash his nuts.

When I visited Phippen's operation that spring, his warehouse was still stuffed with the leftovers of the previous year's almond crop—30 million pounds stored in wooden and plastic bins stacked to the ceiling, and even overflowing into a makeshift storage area he had constructed outside.

Orders assembled for customers in the Middle East and Japan filled white plastic sacks and cardboard cartons, the goods arrayed across wooden shipping pallets. Each was embossed with labels that attested to their prestigious provenance—Travaille & Phippen, a brand known for some of the finest almonds on earth. They were ready to go, except for one crucial detail.

There were not enough ships willing to carry them across the water.

Every week, Phippen peered at a calendar showing confirmed bookings on container ships sailing to points worldwide from the port of Oakland, some sixty-five miles away, on the eastern lip of San Francisco Bay.

Every week, he absorbed all manner of disheartening news. No shipping containers available. No vessel arriving. No space on board.

His almonds were here, in the flatlands of California. His customers were there, on the other side of the ocean. The international shipping industry was failing to bridge the divide.

Phippen walked his darkened storerooms with his face tightened into a look of bitter incredulity.

"My warehouses are already bulging at the seams," he told me. "It scares the crap out of me, because in five months I'm going to get a new crop in the door. There's no timeout in farming."

Phippen's torment stemmed from the same reality that was upending shipping worldwide. Demand for factory goods was fierce. The ocean carriers were continuing to concentrate their vessels on their most lucrative routes—loops between the industrial cities of Asia and the twin ports of Los Angeles and Long Beach.

For American agricultural exporters like Phippen, Oakland was their primary jumping-off point, the gateway for the crops of the Central Valley en route to everywhere else. Yet for the ocean carriers, Oakland had become a place not worth stopping, a speed bump on the path toward record profits.

So the almonds Phippen had harvested the previous year—$19 million worth—sat in his warehouse just as the trees came to life to make more. The days turned to weeks, the weeks to months, with no end in sight.

Two-thirds of the almonds in his storage areas had already been purchased by buyers far away. He would not receive payment until they made it onto a ship.

"These almonds aren't worth shit in the warehouse," he said. "They are worth a lot of money in Dubai."

More than anything, he could not get over the maddening feeling that he was being preyed on by forces beyond his recognition.

"Somebody's screwing with us," he said. "We're getting jacked around here."

For all the agonies that befell American importers during the Great Supply Chain Disruption, exporters had it even worse. Especially those engaged in agriculture.

From wheat growers in North Dakota to soybean producers in Nebraska, shipping crops to customers beyond North America had become perilous and sometimes impossible.

Ordinarily, farming interests enjoyed a symbiotic relationship with multinational retailers, who paid the highest shipping rates, and whose massive volumes of imports ensured a steady supply of shipping containers landing at American ports.

A giant ship would pull into Los Angeles bearing containers full of furniture, clothing, and other imports from Asia. Longshore workers would wield cranes to pluck boxes off the vessels and onto the beds of trucks that would carry them to surrounding warehouses. There, crews would unload their contents before the carriers turned the empty containers over to American farmers.

Some would be sent to the Central Valley—either by truck, or on ships headed up the coast to Oakland—to be filled with almonds, grapes, dairy products, and citrus fruits.

Some containers landed on railcars headed east to the Great Plains and the Midwest, where they were loaded with corn, wheat, and other grains, and carried back to the West Coast. There, cranes hoisted them onto vessels headed across the Pacific, to be turned into bread, cereal, and animal feed in Asia's fast-growing markets.

But the rewards of carrying factory goods from Asia to North America had become so lopsided that they obliterated the economics of the traditional arrangement.

Scoular, one of the largest agricultural exporters in the United States, loaded grains into containers at terminals in Chicago and Kansas City, and then sent them by rail to West Coast ports and onto ships bound for Asia. But with containers scarce, the carriers kept them confined to the coasts. Ships were frequently impossible to book.

A container placed on a railcar headed for Kansas City was a container not available to pick up more factory goods in Ningbo. A ship stopping in Oakland to drop off empty boxes for farm exports was delaying its arrival in Shanghai, where retailers were willing to pay astronomical sums to transport their next load to North America.

So the carriers were increasingly unloading containers in Southern California and then slapping the empties right back on their vessels and returning directly to Asia, without waiting around to load farm products.

Before the pandemic, about 40 percent of all containers leaving the ports of Los Angeles and Long Beach were full of exports. By late 2021, only 30 percent of the boxes leaving Long Beach were loaded with goods, and the rest were empty. Next door in Los Angeles, only 21 percent of the containers on outbound ships were full of goods. On their journeys back to Asia, container vessels were mostly carrying air.

At the same time, the carriers were increasingly bypassing Oakland. Two years earlier, they had canceled planned stops there only 1 percent of the time. Now, they were scrapping a quarter of such stops.

Here was the explanation for the crisis besieging California almond growers. Collectively, they were sitting on 1.1 billion pounds of almonds left from the previous year's harvest. The shipping industry that they depended on to move their product was forsaking them.

"Foreign carriers are being allowed to disrespect us, and we can't do anything about it," Aubrey Bettencourt, president of the Almond Alliance of California, an industry trade group, told me. "We have no recourse."

In Washington, lobbyists working for a host of agricultural interests were pressing that line to growing effect. The carriers were all foreign companies. American farmers were at their mercy.

They had coalesced behind a bill moving through Congress that was supposed to redress such troubles by boosting the authority of the Federal Maritime Commission.

The new chairman of the commission, Dan Maffei—a former Congressman from central New York—was working his former colleagues on Capitol Hill to round up the votes. And the commission was "actively looking to investigate cases where exporters are being pushed around by carriers or, worse, ignored by them," Maffei told me.

No less than the president of the United States had entered the conversation. In his State of the Union address in March 2022, Joe Biden affixed the shipping industry with blame for what had become his primary political vulnerability—anger over rising consumer prices.

"When corporations don't have to compete, their profits go up, your prices go up, and small businesses and family farmers and ranchers go under," Biden declared. "We see it happening with ocean carriers moving goods in and out of America. During the pandemic, these foreign-owned companies raised prices by as much as 1,000 percent and made record profits. Tonight, I'm announcing a crackdown."

The carriers were not accustomed to this sort of denunciation. They were nearly invisible in Washington. Back in their home counties—in capital cities like Beijing, Seoul, and Copenhagen—they maintained gratifying relationships with the officials who kept the subsidies flowing. Worrying about the proclivities of lawmakers from rural America was a new experience.

Public relations representatives for the carriers were typically impossible to find and disinterested in engaging, ignoring media requests for comment.

To the extent that the carriers worried about protecting their public images, they left the work to their industry-wide lobby shop, the imperiously named World Shipping Council. The institution was headed by one of the army of attorneys who quietly dominate dealings in the nation's capital, John Butler, who had been a partner at a boutique firm that specialized in maritime law.

Butler was a skilled practitioner in the Washington art of protecting the client's interest by depicting change as akin to trashing American free enterprise. He had taken to describing the ocean carriers as misunderstood victims of a populist cabal, whose contributions to society were going unappreciated.

Despite the many problems– traffic jams at ports, shortages of truck drivers—the carriers were managing to move more cargo than ever, he noted. The White House and members of Congress were using his industry as props in the shopworn story of giant companies screwing the little guy.

"There's a frustration that really complex problems are getting sound-bited to the point where policymakers aren't dealing with the structural challenges," Butler told me.

The proposed solutions coming out of Washington would worsen the trouble, he warned, while invoking the specter of a terrifying villain: the empowered bureaucrat.

"Do people really believe that having the federal government on the dock with a clipboard saying, 'That box goes on the ship, that one doesn't,' is more efficient and fair than letting the market sort it out?" Butler said.

On the other end of the continent, in Fresno, California, James Blocker was generally predisposed toward siding with people who invoked the sanctity of free markets. He was a proud libertarian, a believer in the frontier spirit of the nation, who bought into the idea that—as Reagan had famously put it—government was the problem, not the solution.

Yet, at the moment, Blocker was dealing with a more prosaic problem: 4 million pounds of almonds waiting for boats to carry them away.

His company, Valley Pride, was among the largest exporters of almonds in California. In a typical week, it shipped fifty containers full of almonds. In recent weeks, the company had struggled to confirm five bookings as the carriers told him they lacked available containers.

He and other almond exporters were getting hit with so-called detention and demurrage charges—fees imposed by the carriers to compensate them for delays in the delivery and return of their containers. These fees could reach several thousand dollars per box after a couple of weeks. And they were being billed even as the carriers themselves canceled their bookings, or failed to provide places for returns, or changed drop-off sites with minimal notice.

This experience had crystallized a sentiment that was alien to Blocker. He was actively calling for regulation.

"I like free enterprise," he said. "I hesitate to get the government

and bureaucracy involved. But we are at the point where we are desperate, and we've run out of options."

The travails of the California almond industry were at once a symbol
of the dysfunction assailing the global supply chain and a testament
to the lunacy of how it had been constructed—that is, an instrument of profit maximization, with all other considerations banished
to the margins.

As environmentalists had long pointed out, growing a single almond consumed nearly a gallon of water. California exported more
than 2 billion pounds of almonds a year. That crop absorbed three
times the volume of water needed to sustain all the homes and businesses in Los Angeles, a city of 4 million people.

The LA metropolitan area represented an audacious experiment
in modern living—a sprawling empire of suburbia dotted by thirsty
lawns and swimming pools, all of it carved into a desert. Ever since
the early twentieth century, when city leaders began diverting river
flow from ranchlands to the north to keep Los Angeles hydrated,
sometimes-violent battles over water had been a permanent feature
of California life. In more recent decades, multiyear droughts and
apocalyptic forest fires had besieged the state, forcing households to
severely curtail their water consumption.

In growing almonds for the world, California was essentially
taking its critically depleted water supply, loading it onto ships, and
sending it thousands of miles away. Here was money dictating the
workings of the supply chain, trumping considerations of climate
change, environmental justice, and common sense.

But the amount of money involved was monumental. Though
California was perhaps best known as the land of Hollywood and
Silicon Valley, its largest industry was agriculture. The state's farm
exports exceeded $20 billion a year. Nearly a fifth of that total was
commanded by the most valuable crop of all: almonds.

Blocker sat in the center of the enterprise. He had grown up in

Clovis, a small town in the Central Valley, on the same parcel of land that his great-grandfather had purchased nearly a century earlier, after arriving from Oklahoma as a refugee from the Dust Bowl. He spent his youth roaming the fringes of Fresno, one of the fastest growing cities in the United States, and a community nurtured by the wealth of the soil.

After college at Fresno State, where he studied agricultural economics, Blocker went to work at Cargill, the agribusiness conglomerate. He started as a commodities trader, managing the firm's positions across North America, and along the way he gained intimate familiarity with the various ways that a crop could be disrupted by unforeseen developments—a shortage of rainfall in the Valley, extreme cold in Canada, an oil shock anywhere.

He started Valley Pride in 2013. The business included a small orchard and a packing plant, but the heart of it was an enormous sales and distribution operation. The company bought almonds from farmers throughout the Valley and exported them around the globe. The previous year, Valley Pride had sold 140 million pounds of almonds while logging revenues of $350 million.

Blocker had to pay farmers for their almonds as soon as they arrived at a port for shipment. But he did not himself get paid until the nuts landed at their final destination. The time between these two events was widening, hitting the company's cash flow. By the time I visited him in Fresno in March 2022, Blocker was tapping a credit line for $8 million to tide the company over.

Tall and sinewy, Blocker was a study in the contrasts of modern agribusiness, a realm that stretched from the commodities pits in Chicago to the dusty farms of California to the bazaars of the Middle East.

He drove a pickup truck, sported a bushy beard, and donned plaid shirts and cowboy boots, holding up his faded jeans with a giant silver rodeo-style belt buckle. He was perfectly at ease crouching in the dirt at his orchard, attending to a leak in the irrigation system. But he spent most of his working hours inside a glassed-in suite at a business complex in Fresno, across the street from a state office building. Deer skulls with antlers—trophies from hunting trips—were mounted above his desk, alongside photos of his wife and three children.

"I'm pretty redneck," he told me. "The big city, the glitz and glamor, doesn't really do much for me."

His business partner, Sunny Toor, hailed from the Indian state of Punjab. He spent much of his time traveling the world developing new markets for California almonds. Their senior vice president for sales, Sorbon Sharifov, grew up in Tajikistan and was fluent in Persian, Russian, Arabic, Tajiki, and English, which aided their push into Central Asia. Others in the office spoke Urdu, Hindi, Spanish, and Serbian.

The previous July, a buyer in Dubai had signed a contract to purchase enough almonds to fill a pair of forty-foot containers. Valley Pride booked passage on a vessel operated by the Mediterranean Shipping Company for a journey from Oakland through the Panama Canal to Malta. There, the containers would be transferred to another ship and carried to Dubai via the Suez Canal.

The vessel was supposed to leave Oakland in October. But the containers did not make it onto a ship until the middle of February. By the time the nuts reached Dubai, their value had dropped by $50,000. The buyer claimed that the initial contract had expired and demanded a discount.

Toor and Sharifov spent most of February in Dubai, feting customers at high-rise restaurants to try to mollify their anger over delays while fending off demands for price breaks. But the shipping problems continued, and their revenues were coming in at half the previous year's.

"We're in a panic situation," Blocker said.

It was just after eight o'clock on a Tuesday morning. He and Toor were preparing to dial into a conference call with the sort of professional whose services they had never imagined needing: a Washington lobbyist.

Like nearly every other practitioner of his craft, Peter Friedmann had previously worked as a staffer on Capitol Hill. For more than three decades after, he had run the Agriculture Transport Coalition, an advocacy shop directed at helping farming interests export their crops.

With the studied patience of a professional wise man, Friedmann listened to Blocker and Toor lament the infuriating state of shipping. He assured them that their troubles were near-universal.

Hay farmers in the Pacific Northwest, who sent bales to Asia, were not even bothering to cut their crops. What was the point of buying fuel for their machinery when there was no space on ships, and when domestic hay prices were plunging because of the resulting glut? Cherry farmers were being decimated by low prices as exports got stuck in North America.

What the carriers were doing was at once horrible and rational, Friedmann explained. Bringing in imports from Asia brought ten times the profit as moving almonds in the other direction.

Ninety percent of Valley Pride's almonds went out of Oakland. Blocker wondered if he should explore other routes. He had contemplated shipping from the port of Savannah, Georgia, where freight was moving more easily. But getting his shipments across the country by rail could take two weeks.

He had just returned from a reconnaissance trip to Houston, where he had learned that containers were abundant. He had lined up warehouse space there in preparation for shipping out of the Gulf of Mexico.

Trucking almonds to Houston would add about $2,800 to every shipment. That seemed worth it given the situation. But then Blocker learned that ships would not take containers out of Houston for three more months, and only with premium charges reaching $5,200 per box, double the usual cost of reaching Dubai.

Friedmann was focused on persuading the carriers to pick up more shipments in Oakland. He was huddling with Biden administration officials to apply pressure. Still, he tempered expectations.

Agriculture was not a core constituency for this president. The Federal Maritime Commission had displayed minimal willingness to use what limited authority it possessed.

In an ideal world, he added, Biden would summon the heads of the ocean carriers to the White House and demand that they fix these problems or prepare for expensive inconvenience unleashed from on high. But that was not going to happen. Despite their usual symbiotic shipping roles, agriculture and retail were then locked in a zero-sum competition for space on ships. Any directive that the carriers pick up boxes loaded with farm produce was effectively an order to delay loading those same containers with cargo destined for Amazon and Walmart.

"It will slow the supply chain for the imports," Friedmann said.

Blocker took this in like a drowning man hearing that the life preservers had been purchased en masse by Jeff Bezos.

"I feel helpless," he said. "We can talk about this, and it's just like, it's intellectual masturbation. It does feel like it has to be from the top, the highest powers of the country."

One of his logistics staff popped her head into his office bearing a rare piece of good news. She had managed to book five containers from Oakland to Dubai, departing within two weeks.

Blocker was pleased, but not wildly so.

"It's a piss in a big ocean," he said. "We're in the land of the forgotten."

The next day, Blocker got in his pickup and drove 110 miles north, to the unremarkable town of Manteca, to visit his most important customer—Scott Phippen.

Blocker revered Phippen like a second father. The previous year, one-fifth of Valley Pride's sales had come from Phippen's operation. Blocker was handling 90 percent of Phippen's exports to the Middle East, and half of his business in Europe.

"My job right now is to take care of Scott," Blocker said. "We've attached our cart to his horse."

Gray clouds hung low over the Valley as his truck threaded the expanse of orchards—newly planted almond trees, bushy orange trees wrapped in green plastic netting, then a grid of pistachio trees whose limbs extended in stops and starts, weaving mischievously skyward like a garden fit for a haunted house.

The towns along the way displayed the centrality of farming in the Central Valley. He rolled past a well and pump company, a factory that made discs for tractors, seed distributors, and insurance agents that catered to farms. He passed flatbed trucks carrying irrigation gear, fertilizer, and machinery for tearing the husks off nuts.

Phippen greeted him in a conference room inside a utilitarian building just off the road.

A third-generation Valley farmer, Phippen's grandfather had arrived from his native Holland a century earlier, after a brief stop in Iowa. Phippen had grown up driving a tractor during the harvest. Now sixty-seven, his sunburned cheeks and calloused hands attested to a life spent outdoors on his 2,500-acre orchard. He attended to a shelling plant that squeezed almonds out of their natural armor, and a processing facility that pounded nuts into powdered form. Discarded shells were piled into a mountain near the back of his yard. Phippen sold them off as feed to surrounding dairy farms.

"This is a big toy box for me," Phippen said. "There isn't anything in this business that I haven't done personally."

Blocker had known him for a decade, ever since he cold-called Phippen, putting forward a plan to sell his lesser-known varieties. The pitch had impressed the older man, who tended to be curt and dismissive of salespeople. Phippen did not need an agent to handle his premium wares. But what about his oddball varieties, almonds with names like Aldrich and Winters? Phippen was selling them generically as "California almonds," with standard prices to match. Blocker proposed marketing them in the same way that wine merchants elevated unknown grapes—by branding them as sexy new varietals with distinctive identities, rather than simply turning them into cheap table wines.

Phippen was game to try. When Blocker succeeded, a lucrative partnership was born.

Phippen could be meticulous to the point of farce. As he wandered his warehouse, he noticed one giant bag of almonds sagging ever so slightly to the left. He stopped and frowned.

"I want them standing up straight," he said.

But his attention to detail was central to the enterprise. Much of the almond processing business amounted to shaking trees, running heavy equipment to scoop up the resulting rain of nuts, and feeding the harvest into a diabolical array of machinery that separated out the valuable bits from the detritus—bark, pebbles, glass shards, and earth. The more rigorous the process, the more the nuts were worth.

"I'm a control freak," Phippen said, not by way of apology but as a point of pride. "When I have to count on other people to do things for me, I just don't trust it will get done right."

Blocker had won his confidence. The two men spoke every day, often more than once.

But what they had been talking about lately, to the exclusion of virtually all else, was how to get almonds onto ships.

In a typical month, Phippen required about a hundred containers to handle his flow of exports. In January 2022, he had managed to ship sixty-six. The next month, fifty-five. Recent weeks had produced a more modest sum: zero.

He was by then sitting on enough leftover almonds to fill 678 containers.

"You get to the point where you're defeated before you even get to work," he said. "It's a sick feeling."

His inability to ship product had turned even virtues into liabilities. Favorable weather the previous year had yielded a robust crop, but that now meant that his storage anxieties were even greater. Some of his stock was piled up outside, with only tarps to protect the almonds from the elements.

He had spent $1.5 million to construct a new warehouse that was supposed to prevent a repeat. He shelled out $820,000 to buy three thousand more bins. He was readying plans to build yet another warehouse, spending another $700,000.

The only way to generate the cash flow to finance this was to pay the ocean carriers whatever they demanded.

"How much do they want?" he said. "I'm ready to take out my checkbook, and still, we can't get bookings."

Blocker shrugged. "You can agree to pay the ransom," he said, "but it doesn't get your load on board."

CHAPTER **9**

"I THINK I'VE HEARD OF THEM."

The New Sheriff on the Docks

Dan Maffei did not grow up dreaming of one day running the Federal Maritime Commission.

Raised amid the maple trees of Syracuse, New York, he never even saw the ocean until he was eleven. As a child, he was fascinated by science fiction, imagining a life as an astronomer. After college, he worked as a local television reporter before jumping into politics, overseeing communications for powerful Democratic lawmakers in Washington, and eventually winning his own seat in the House of Representatives.

But after he lost his reelection bid in 2014, Maffei, then in his midforties, was casting about for the next thing to do. He was not interested in the usual route for deposed members of Congress—lobbying former colleagues.

Seeking counsel from friends in the Obama administration, he heard there was likely to be an opening for a seat on a federal commission.

Maffei's ears pricked up. The Consumer Product Safety Commission? That could be interesting.

No, they told him. The Federal Maritime Commission.

"I said, 'Well, okay, I think I've heard of them,'" Maffei recounted. "'I'm already ahead of the game.'"

In the summer of 2016, he took a seat on the five-member commission. Trump reappointed him. When Biden took office, he elevated Maffei to the chairmanship.

It seemed reasonable to expect that Maffei would be left to dig into the minutiae of the shipping business, far from the histrionics of whatever issue might be consuming the nation's capital. Since its creation in 1961, tranquility and obscurity had characterized life at the commission.

But that changed dramatically with the arrival of the pandemic and the Great Supply Chain Disruption. Suddenly, the commission was thrust into the center of a high-stakes political issue with enormous economic consequences.

Congress reacted to the anguished accounts of agricultural exporters like Phippen and importers like Walker by coalescing around a bill engineered to bolster the commission's authority, the Ocean Shipping Reform Act. In a rare bipartisan showing, lawmakers in both chambers championed the legislation as redress for the abuses of the ocean carriers.

As Biden signed the bill into law in June 2022, he portrayed it as follow-through on his promises to bring the ocean carriers to heel. It would "put a stop to shipping companies taking advantage of American families, farmers, ranchers and businesses," Biden declared. "This bill is going to help bring down inflation."

But even the president was not entirely clear on the particulars of the body he was counting on to administer justice.

"What this bill does is extend the authority of something the federal government calls—" Biden paused, briefly unable to summon the words to complete his sentence "—the Federal Maritime Commission. Which a lot of people don't know exists, but it's important."

The commission would force the ocean carriers to put agricultural exports on their ships. It would relieve importers from outrageous fees. It would ensure that carriers honored their contractual obligations.

To underscore all this, Biden called Maffei up to the podium to stand behind him as he affixed his signature to the legislation.

Maffei leaned forward and whispered into the president's ear, thanking him for not having hesitated in appointing a landlubber to the maritime commission. Biden chuckled.

Then, with the stroke of his pen, the president tasked the commission with defusing the most alarming crisis of his term: soaring consumer prices.

Maffei had become a Washington archetype—an obscure figure virtually unknown outside the Beltway who was suddenly thrust into a central role in a critical mission of international consequence. Here was the guy who was supposed to break the monopolistic excesses of the shipping carriers.

It seemed like a lopsided battle.

The ocean carriers had tens of thousands of ships, connections to governments in their home countries, offices around the world, and hundreds of billions of dollars in profits. They were annually transporting upward of $14 trillion worth of products.

The Federal Maritime Commission was armed with a staff of only 120 people. Its annual budget was a mere $32 million—roughly the profits secured by the carriers every hour.

In contrast to the colonnaded fortresses housing the agencies that dominated official Washington—the Treasury, the State Department—the commission occupied two floors of a nondescript office building tucked on a lifeless block, alongside a prep school football field. The first time I arrived there, the security guards were unfamiliar with the commission, and unclear on how to reach it. One of Maffei's staff had to come down to the lobby to find me. We rode up in a service elevator with padded walls.

In theory, the commission was the place that a shipper would go seeking remedy for ill treatment by an ocean carrier. But that hardly ever happened. The commission's authority was limited, and its members traditionally displayed deference to the carriers.

"They became hostage to the industry," said Peter Friedmann, the lobbyist who represented agricultural exporters.

A shipper who suffered a bad experience generally kept it to themselves, lest they suffer retribution without justice.

In the rare event that a company filed a formal complaint, the case was adjudicated by an administrative law judge, with the commission involved only on appeals. The body had scant power to levy fines or compel action. At least, that was how its attorneys tended to construe the law.

It did not seem coincidental that the commission had frequently been headed by people with ties to the shipping industry, such as Elaine Chao, whose husband was Senate Republican leader Mitch McConnell, and whose family controlled significant shipping interests in China.

Maffei broke from this mode. He was an outsider with no history in the industry. He took on the assignment with earnest curiosity, reading maritime histories, visiting ports, and filling his capacious office suite with the paraphernalia of the sea—antique maritime clocks, old maps, models of giant container ships.

But there was one complicating factor in the notion that Maffei would take on the dominance of the ocean carriers. He was not especially troubled by monopoly power.

In contrast to the fiery rhetoric emanating from the White House, the chairman of the maritime commission favored more measured, less confrontational words. Biden had seized on the ocean carriers as a primary villain in the tale of inflation. Maffei was a man conditioned to avoid narratives of perfidy. He distilled conflict down to competing interests that required balance, finesse, and understanding.

A Democrat from a congressional district animated by a sharp partisan divide, he presented himself as a centrist, a pragmatist, and an all-around regular person. He liked *Star Trek*. He had once tried and failed to shotgun a beer with the television comedian Stephen Colbert, on a segment introducing his little-known community to the viewing world. He was an easy guy to get along with, a good listener, confident but not arrogant.

"I'm just extraordinarily ordinary in some ways," he once told an interviewer for his local paper. "I grew up with two middle-class parents who got divorced. I went to public school. I wasn't the star of the lacrosse team."

But he had earned no fewer than three Ivy League degrees— Brown undergrad, Columbia Journalism School, and Harvard's John F. Kennedy School of Government. He had worked for some of the more famously cerebral figures in the history of the Senate, Daniel Patrick Moynihan of New York and Bill Bradley from New Jersey. He possessed an analytical bent that tended toward shades of gray while eschewing the more colorful vernacular of political denunciation.

In contrast to the president who had made him chairman, Maffei rejected the idea that sharp price increases imposed by the ocean carriers reflected their dominant hold on the market. In his telling, the shipping industry could not be viewed through a traditional antitrust lens because it was a peculiar business, one guided more by national interests than the profit motive. And this had long been a benefit to American importers and exporters. They had helped themselves to cheap shipping rates.

"The status quo wasn't a problem for two decades," Maffei told me.

He worried that if Washington took too hard a line, the carriers might reduce their American ports of calls, leading to higher costs for transport, and raising the price of consumer goods. They could ship into Canada and Mexico, and then truck goods into the United States.

"At the end of the day," Maffei said, "they don't have to serve us."

This was a stark departure from the way in which American officials traditionally described their power: as an extension of the nation's economic clout. The United States possessed the largest marketplace on earth. That made it too big for any multinational company to ignore, and Washington generally behaved as if this fact conferred the authority—if not the moral right—to dictate the terms of trade.

But Maffei's gut-level aversion to confrontation reflected the reality that American businesses and consumers had grown dependent on a supply chain dominated by foreign ocean carriers. The United States was addicted to a commodity controlled by Chinese state-owned companies and other national concerns around the world. He feared pushing them too hard.

The new law directed the commission to bulk up its enforcement capacity while creating systems that would make it easier for aggrieved shippers to file formal complaints. It increased the agency's funding, lifting it by half over the subsequent three years.

But Maffei saw the details of the reform as less important than its very existence. Even before the passage of the new law, deliberations in Congress had motivated the carriers to begin picking up more containers in Oakland, a boon to agricultural exporters.

"Once they knew it was going to happen, we had movement," Maffei said at a gathering for farming interests.

The process yielding the legislation had conveyed a key message to the carriers. They had to fix the problems themselves, quieting complaints, or risk intervention from on high.

"Deterrence is what it's about," Maffei told me. "On a day-to-day basis, we're too small an agency. We're never going to catch every instance."

We were having breakfast at an Irish-themed pub a couple of blocks from his office in the summer of 2022. Maffei wore a New York Yankees baseball cap and a brown polo shirt embossed with the maritime commission's logo. He was carrying a cup of coffee, the first of an uncountable run of caffeinated beverages that would fill out his day. When the waiter approached, he immediately ordered another cup along with the "full country breakfast," though he pointedly excised the blood sausage. "But I want the bangers," he emphasized.

He was clearly still adjusting to life outside the Capitol. "I have a lot of nightmares about Congress," he said. "There's definitely a past still haunting me." He had recently had a dream in which he had gotten a peek at a Republican strategy document. He relayed to House Speaker Nancy Pelosi what the other party was going to do. Then they did something else. "Pelosi was mad at me," Maffei said.

Still, his time in Congress was invaluable to the job of running the commission. He had helped build the coalition that produced the ocean reform bill, which had become his to-do list. He wandered the bowels of the Capitol like a native, stopping regularly to greet staffers and lawmakers.

But he clearly had to expend effort in managing the expectations of the White House as the political stakes of inflation grew. The president was publicly demanding a "crackdown." Maffei was seeking to satisfy such demands in spirit without further disrupting the supply chain.

Another maritime commissioner, Rebecca Dye, a Republican, had recently produced a voluminous report detailing the impacts of the pandemic on the ocean shipping industry. Dye was widely respected for her institutional memory. Her report was full of data, providing a catalogue of trouble. But within the political ranks, her sixty-five-page document could be boiled down to one money quote near the beginning. Though shipping prices were "disturbingly high by historical measures" they were also "the product of the market forces of supply and demand." In short, no foul play.

The White House must have hated that report, I suggested.

"'Hate' is a strong word," Maffei replied.

Over the course of breakfast, he patiently took apart the administration's contention that American consumers were paying more for shoes and furniture and clothing because of years of consolidation in the shipping industry. The carriers had lost money for years amid an oversupply of ships. With capacity now tight, they were making up for lost time.

But wasn't the inequality arising from this chaos a problem? Amazon and Walmart could afford to charter their own ships—a huge advantage. The supply chain disruption and high shipping rates seemed certain to diminish competition, I suggested, leaving the largest companies in a stronger position. Wasn't this a victory for companies that had already won? What would happen to niche players like Hagan Walker? What would become of the almond farmers in the Central Valley?

Maffei regarded me as he would someone complaining about the heat in August. The rewards of scale were a simple fact of life.

"The small- and medium-size folks are boxed out," he said. "That's capitalism."

He analogized this to the disappearance of Mom 'n' Pop burger shops in American communities, as McDonald's restaurants and Burger King outlets proliferated. "Does that mean there's no competition?" he said. "No, they compete like crazy. At the end of the day, there's a reason that companies try to get bigger."

Maffei was affirming the central idea that had allowed decades of consolidation, not just in shipping but in rail, trucking, retail, meatpacking, telecommunications, and seemingly every other crevice of American economic life. He was accepting that the forces shaping the supply chain—shareholder imperatives, the pursuit of scale, the profit-making efficiencies of Just in Time—were an ineluctable part of American reality. His job was to keep the gears turning, not refashion the machine.

But how did he square this with the language coming out of the White House? The president and his staff were describing an organized rip-off of the American consumer. Maffei was effectively softening a criminal indictment into a civil complaint.

"There is a rip-off," he said. "But explaining where the rip-off is doesn't fit easily into a quick speech. It's not in the rates. It's in these fees."

He was talking about the detention and demurrage charges that carriers were imposing. He meant the premium levies and congestion fees. It was not market power that was to blame for the turmoil at the ports, he said, but rather a bit of opportunistic sleight of hand by companies that were used to having no cop patrolling the beat.

This was a much narrower case against the ocean carriers than the one pressed by Biden. It was hard to see how much would change. Maffei would take on the industry, but in his own fashion, through tact and diplomacy.

Just after 9 a.m., Maffei strode into his suite at the commission's office. He changed into a dark blue suit with a necktie embossed with an anchor pattern. Then he sat down to go over the statement he planned to make at the commission's monthly meeting, which was to be held later in the morning.

The meeting was dedicated to the commission's plans to implement the Ocean Shipping Reform Act. The chairman was especially eager to convey the message that enforcement was being intensified.

Just before ten, he took his place on a wooden dais inside the commission's official chambers and looked out at the gallery—three dozen chairs, sparsely filled with lawyers and lobbyists representing the ocean carriers and their customers. If shippers like Phippen and Walker were going to get any relief for their troubles, this was the place it was supposed to happen.

The meeting was a fiasco. Maffei banged his gavel to call the meeting to order, only to be told that people watching the proceedings online could not hear the audio. The tech people were trying to fix it, but no one knew how long this might take.

Nearly two hours later, the system was still down.

"We've been trying to get the hearing room fixed," he said. "You can tell it's kind of old. Everything has to be, to a certain extent, jury-rigged."

He gave up on an in-person meeting, opting to convene his fellow commissioners on a videoconferencing platform. Back in his office, he propped his laptop on a thick volume of maritime regulations. An aide moved an American flag from near the window to the back of his office to get it in the frame. He wielded a coffee mug as a gavel.

"The commission will come to order," he said, as some of his colleagues froze and others gestured that they could not hear.

"This is going to be interesting," he muttered.

His staff presented their plans to recruit more enforcement staff. The commission would make publicly available the data that the carriers were now required to provide, detailing their volumes of exports and imports from port to port. The agency would gather reports of noncompliance and pursue cases.

Then Maffei leaned into his laptop and underscored the stiffened posture of his agency.

"This is the law of the land," he said. "If you have a complaint about it, we can direct you to the Congress or the White House."

He broke for lunch with his staff in a conference room—chicken brought in from a nearby Peruvian restaurant, washed down with Coke Zero.

He met behind closed doors with a delegation from a major European carrier.

Late in the day, back in his office, he called the head of the mighty Port Authority of New York and New Jersey, Bethann Rooney, who briefed him on the mayhem besieging her corner of the world. They were running out of places to stash containers, because the docks were choked with stacks of empties—more than two hundred thousand of them. The carriers were not sending enough ships to collect them. Instead, they were deploying their fleets to cash in on the bonanza on their trans-Pacific routes.

Maffei listened to Rooney's account while slumped in a wingback chair, facing an oil painting by a seventeenth-century Dutch artist depicting two sailboats caught in frothy surf near menacing rocks.

Everything was backed up, Rooney reported. With space tight at the port, local truck drivers were being denied appointments to return empty containers. And still the carriers were sticking them with fees for the late return of their boxes. The heads of local trucking companies were apoplectic.

This was the variety of rip-off that provoked Maffei's ire.

Would it be helpful for him to tour the port? he asked. He could meet with trucking companies, put out a statement of concern, and underscore the new atmosphere for the carriers.

"A brief, symbolic visit," he said.

Yes, Rooney replied. It couldn't hurt.

The following week, Maffei arrived at the Port of Newark.

Nearly seven decades had passed since Malcom McLean embarked on his maiden container voyage from these very shores. The terminals clustered on the land had become a linchpin of the global supply chain.

Tractor-trailers rumbled through every second, carrying containers in and out of the gates of the port. Cranes stretched skyward, plucking boxes off incoming vessels and adding them to the stacks.

Inside a port administration building, Maffei walked a slow turn around a long conference room table, greeting the heads of a dozen local trucking companies. When they sat down, each presented the chairman with a similar set of grievances.

They would call the carriers seeking to return empty containers, only to be told that no appointments were available, or the drop-off locations were full. Then they would receive bills reaching $150 per box in daily detention charges for failing to return their containers. Until they paid, the carriers would refuse to release other cargo they were carrying.

"We call it ransom," one said.

A man named Tom Heimgartner, who chaired a grouping of local trucking firms, the Association of Bi-State Motor Carriers, implored Maffei to use his powers to force the carriers to clear out the backlog of empties.

"Our port is gridlocked," he said. "It's an emergency. We need something done here."

Maffei listened intently and jotted notes in a pocket-size journal.

The trucking company leaders beseeched him to declare a moratorium on detention and demurrage charges until the carriers took away enough empty boxes to relieve the congestion.

But the commission lacked the authority to do that, Maffei said. The carriers would have to agree to such a policy voluntarily, though he could certainly apply pressure to bring them around.

But there was one potentially useful avenue to explore, he suggested. The carriers appeared to be violating the shipping reform act by effectively forcing the trucking companies to store their containers without compensation. Because they could not drop off empties, the trucking companies were stuck with the boxes. They were stashing them in their yards. The commission could perhaps order the carriers to pay the trucking companies for the storage of their boxes. That would give them an incentive to clear out the backlog.

The truckers were excited by this possibility. But it required them to take a highly unusual step. They would have to lodge formal complaints at the commission. That meant picking a fight with the ocean carriers, on whom their livelihoods depended.

"It sounds like they are treating you like such dirt," Maffei said. "I'm not sure you have anything to lose."

Jacob Weiss had seen what could happen when someone had the audacity to challenge a shipping carrier. It was like a minnow picking a fight with a whale.

His company, OJ Commerce, had encountered the same familiar constellation of troubles suffered by thousands of other shippers as he sought to move furniture from factories in Asia to his customers in the United States. His carrier, Hamburg Süd, a division of Maersk, was refusing to accept many of his containers at his contracted rate, forcing him to pay far more in the spot market.

But unlike other shippers, Weiss opted to fight.

In April 2021, he instructed his lawyers to fire off a menacing letter, warning the carrier that it had to "immediately honor" his contract or invite a formal complaint at the Federal Maritime Commission.

The result of the letter was swift and decisive, though not in the way that Weiss envisioned: Hamburg Süd immediately cut off talks about renewing his company's contract for the following year. "We should not engage in any renewal discussion with customer in light of potential litigation," a senior vice president at Hamburg Süd North America, Juergen Pump, wrote in an internal email that OJ Commerce filed as part of its case at the maritime commission. "I would also not provide them with space under the existing contract."

This seemed a blatant act of retaliation, a course expressly barred under the shipping reform act.

Maffei would not discuss the specifics of the case, but—speaking generally—the chairman condemned retaliation as a fundamental attack on the notion of justice.

"The clear intent of Congress, and my intent as well, is to come down as hard as possible on any kind of retaliation," he told me. "It undermines the entire system of enforcement."

But those words could not compensate for the reality that the system of enforcement was vulnerable to myriad forms of manipulation. In this case, Hamburg Süd effectively ran out the clock to avoid redress.

For nearly a year after filing its case, OJ Commerce sought to use the discovery phase of the maritime commission's legal proceedings to gain access to the carrier's pricing data. It was trying to establish how much money the carrier was making by selling its space on vessels to other shippers.

An administrative law judge assigned to the case, Erin M. Wirth, set a deadline for discovery, allowing the carrier and OJ Commerce to request records and depose witnesses. Twice, she extended the deadline while ordering Hamburg Süd to hand over records and make its executives available. But the carrier repeatedly ignored her directives, even as she threatened sanctions.

At one point, the carrier produced a witness who immediately asserted that he knew nothing about the case. At another, the carrier agreed to make a key executive available, but only on a single day: the Jewish holiday of Yom Kippur. An orthodox Jew, Weiss could not work on that day.

By October 2022, the clock had expired on discovery. The judge directed OJ Commerce to do the best it could in briefing the issues before her ruling, even without the data it had sought to make its case.

Six months later, the Maersk subsidiary was still stalling the proceedings with arguments over what material should be deemed confidential. Wirth had again extended the case while withholding judgment.

Weiss's experience exemplified why shippers, despite their excruciating dealings with the ocean carriers, generally kept their frustrations to themselves. They rarely complained to the press, let alone filed legal challenges at the Federal Maritime Commission, because they understood that they ultimately had little recourse. Concepts like due process and fair hearing were quaint notions alongside the naked power of the shipping behemoths.

And this was the central cause for skepticism as the Biden administration trained its sights on the industry, and as Maffei's commission crafted new rules to regulate it. However those rules turned out, the shipping carriers would retain an overwhelming advantage over their customers. They had lobbyists, lawyers, and the fleets themselves. They had time to drag out legal proceedings. Whereas farmers like Phippen and retailers like Weiss had products to move right now, which put them at the mercy of the conglomerates that controlled the ships.

It was hard to envision how that basic reality would be altered.

CHAPTER 10

"EVERYTHING IS OUT OF WHACK."

Floating in Purgatory

Hagan Walker had no time to contemplate the niceties of maritime contracts. He needed his stuff on land, immediately.

By the time the *Maersk Emden* arrived in the waters off Long Beach just after six o'clock in the evening on October 9, 2021, the Christmas season was less than three months away. The pressure to make good on his *Sesame Street* order was mounting. Yet his container was still stacked on the ship, which was floating aimlessly out in the Pacific, awaiting an open berth.

The wait seemed certain to take days. Perhaps even weeks. Seeking an estimated time of arrival from his shipping agent rivaled inquiring about the meaning of life. The havoc consuming Southern California's ports was only intensifying.

This was a problem that went well beyond the fate of Glo's *Sesame Street* figurines. Collectively, the twin ports of Los Angeles and Long Beach were the entry point for two-fifths of all American imports arriving by container ship, a flow of goods that was in the midst of an unprecedented surge.

From factories to restaurants to retailers, businesses were ordering extra supplies and parts, cognizant that delays in delivery were a certainty. Long ruled by the balance sheet–enhancing powers of Just in Time, companies were suddenly padding their orders with surplus.

Here was a feedback loop of increasing demand. The more that businesses raced to fill their warehouses, the greater the strain on the supply chain, increasing the odds of delay. All of which generated an even greater imperative to order more stuff now.

The volume of imports moving through the Southern California ports would increase by 16 percent over the course of the year.

With the docks overwhelmed, the ports were forcing ships to idle out in the ocean, waiting their turn to unload. On the day the *Maersk Emden* reached the vicinity, the line exceeded fifty vessels—up from zero before the pandemic.

For the first six days, the *Emden* did not even have a dedicated place to park. It ran a slow, looping course in the waters off the port, before anchoring in formation with nine other vessels, roughly three miles off the coast. There it sat for another ten days, consigned to serving as a floating warehouse.

Dead ahead of the *Emden* sat the *Maersk Essex*, which had arrived from the port of Xiamen, in southern China. To the rear was the *Kassos*, a Liberian flagged oil tanker just in from Vancouver, Canada, on its usual loop between the Middle East and North America. Off the *Emden*'s starboard sat the *Cosco Nagoya*, a Panamanian flagged container ship that had arrived from Ningbo.

Stuck on board the ship, the two dozen crew—mostly Filipinos and Indians—bided their time playing basketball and Ping-Pong, watching movies, singing karaoke, and using patchy Wi-Fi links to text with spouses and children on the other side of the world.

"It's very sad and lonely away from home," said Alejo Cuyo Jr., a Filipino crew member who worked on the *Emden*'s sister ship, the *Maersk Essen*, which ran between Asia and the West Coast of the United States.

Raised in a village that lacked plumbing and electricity, he recalled hauling water with a bucket, walking more than a mile to school, and doing his homework by the light of a kerosene lamp. He took pride in earning enough—about $2,000 a month—to maintain his own family in a solid home with modern conveniences, about 170 miles north of Manila.

But his time away sometimes lasted for as long as six months. He felt himself sinking into despondency as the vessel pitched, as he subsisted on frozen foods, and as he wondered when he would next see his wife and three children.

"Life at sea is very difficult," he said.

Walker had every reason to be concerned about the fate of his order. After arriving in the waters off the ports of Los Angeles and Long Beach, the average ship was then waiting nearly two weeks before it could pull up to the dock. By the end of November, the wait extended to nearly three weeks.

Even after ships finally reached a berth, delays continued to bedevil loading and unloading, given that a rotating cast of dockworkers and truck drivers were stuck in quarantine. The ports were operating at about two-thirds of their capacity.

As the *Maersk Emden* bobbed at anchor, the White House outlined plans to unclog Southern California's ports by having them operate around the clock—twenty-four hours a day, seven days a week.

"Today's announcement has the potential to be a game changer," President Biden declared. Soon, he promised, Americans would be relieved of their worries about shortages afflicting products from "toasters to sneakers to bicycles to bedroom furniture."

This was a stretch. The president did not possess the power to restore order to the interlinked transportation industries that moved products around the world.

Biden could and did extract promises from the heads of the ports of Los Angeles and Long Beach to deploy more dockworkers, who were delighted to gain more paid hours. He secured public commitments from major carriers to move more products. But he had no authority to hire more truck drivers at a time when trucking companies were decrying shortages.

Nor did he have sway over warehouses, which proved the ultimate impediment to his aims. In areas near the nation's largest ports—in Southern California and in the New York area—warehouse vacancy rates were at 1 percent and below, meaning there was frequently nowhere to stash incoming goods. This was why retailers were leaving their boxes uncollected at ports, amplifying congestions. Warehouse operators typically lacked the incentive to stay open at all hours. That would require that they hire more security staff along with more people to move products across the floor, none of which made economic sense. Retailers were not adopting 24/7 schedules, limiting their desire for extra shipments.

No presidential decree could summon more chassis, the wheeled carriages pulled by trucks to transport containers, whose production was dominated by Chinese companies. And Biden's words could not eliminate the stacks of empty containers clogging many major ports. These two problems reinforced one another.

Containers were stuck outside warehouses that were already stuffed with goods. The boxes typically sat on the chassis on which they arrived. Most warehouses lacked equipment needed to lift a container off a chassis. So nearly every container that sat in a warehouse yard was tying up a chassis. Trucks seeking to haul empty containers from ports often failed to turn up for their appointments because they could not locate a chassis. Instead, the boxes sat uncollected at the docks, jamming up the works and extending delays up and down the supply chain.

But the biggest reason to doubt Biden's bold pronouncement was the simple fact that the people who actually had power over the ports were benefiting from the continued chaos, challenging their motivation to ease the congestion. The municipalities of Los Angeles and Long Beach were just the landlords to the ports. They collected lease fees from thirteen container terminals that ran the cranes and the forklifts, overseeing the loading and unloading of vessels. The terminals were almost exclusively controlled by the ocean carriers.

This presented an enormous conflict of interest.

For the carriers, the spectacle of dozens of ships marooned off the coast amounted to a source of monetizable market anxiety. It inspired people like Hagan Walker and James Blocker to hand over whatever sums were required to get their boxes onto ships. It created a plausible justification for exponential increases in price, muting uncomfortable conversation about the role of monopoly power. It allowed the carriers to justify their otherwise-insane prices as the simple product of supply and demand.

The carriers strenuously rejected talk that they were profiting from the disruption at the ports. The traffic jams were actually costing them money, they insisted.

"Freight rates have been impacted by the global Covid-19 recovery and the demand outpacing supply, not simply as a result of port congestion," a Maersk spokesman, Tom Boyd, said in an emailed statement. "Ships at anchor are not productive, nor are they earning revenue against a backdrop of large fixed costs."

But Maersk alone—which operated APM, one of the largest terminals at the Port of Los Angeles—somehow managed to harvest nearly $62 billion in revenue in 2021, an increase of more than half compared to the previous year. That was enough to hand shareholders $6.5 billion in dividends.

"The third quarter was the best quarter in the history of the company," the carrier's chief executive officer, Soren Skou, boasted to investors a few weeks after the *Maersk Emden* arrived in Long Beach.

The carriers did not create the congestion at the ports. Supply and demand indeed largely explained that part. But there were solid reasons to doubt their motivation to ease the traffic jams as quickly as possible.

Skou and his fellow executives running the other terminals at the ports in Southern California could have boosted their capacity to move cargo by deploying more dockworkers, adding shifts to the calendar. But such seemingly logical actions went against the Just in Time ethos that ruled the supply chain. Instead, they hired more casual workers—the part-timers who lacked the protections and wage scales of full-timers. They pushed to install more robots.

"Every terminal should be working that third shift," said Jesse Lopez, secretary and treasurer of the International Longshore and Warehouse Union Local 13, the chapter that represented laborers in Long Beach and Los Angeles. "We are willing to work, and we are ready to work."

Dockworkers routinely complained that loading and unloading was perpetually delayed by the failure of aging and shoddy equipment. Despite the "best quarter in the history of the company," Maersk was reluctant to upgrade trucks and other machinery that was constantly breaking down.

"It doesn't seem like there's much incentive to move the cargo faster," said Jaime Hipsher, a dockworker who drove heavy equipment at APM, Maersk's terminal at the Port of Los Angeles.

On a recent shift, she had been moving containers between stacks on the dock and a rail yard. Only five minutes in, her vehicle broke down. She switched to a backup truck. An hour later, that one broke down, too.

For Maersk and other shipping terminal operators, moving faster to ease congestion meant eradicating the conditions that were allowing them to log record profits.

By the middle of 2022, Maersk was celebrating another record quarter, with revenue reaching $21.7 billion between April and June—an increase of 52 percent. Skou, the carrier's chief executive, was promising investors that Maersk would maintain tight conditions to keep shipping prices high.

"We will provide the capacity that our customers need," he said, "but we will not sell all the capacity that we have unless there's demand for it."

On a sunny afternoon in March 2022, I boarded a boat with the head of the Port of Los Angeles, Gene Seroka, as we surveyed the congestion on the docks—the cranes operating full-on, the towering stacks of containers holding goods that were supposed to be somewhere else.

"The entire supply chain has been crooked for over two years now," Seroka said. "It's broken, it's indifferent. Are there people who profit off of the inefficiencies? Sure."

Like Maffei, Seroka was another previously anonymous official who was suddenly on every news channel, dressed like a politician in a dark suit and shiny tie, huddling with no less than the president of the United States, and solemnly vowing to restore normalcy to the supply chain.

The ships arrayed across the water, hulking and motionless, afforded an unusual glimpse of the typically invisible machinations of the modern world. Television crews could not get enough of the spectacle. As far away as Des Moines, ordinary people suddenly knew how many ships were in the queue off Los Angeles on any given day, recounting the number like the temperature in a record heat wave or the wind speed in a lethal hurricane. Here was a gauge that explained why children were going back to school without proper notebooks, why couples were compromising on the cabinets for the new kitchen, why repairing a car could take months.

Similar scenes were filling out horizons off scores of other ports. More than fifty vessels were stuck waiting near the docks of Shenzhen and Yantian in China. At least forty vessels were floating off the port of Ningbo. And on the other end of the United States, more than twenty container vessels were anchored near the port of Savannah, Georgia, floating as far as seventeen miles off the coast in the Atlantic.

Savannah was best known for its languid vibe, its modern Southern bistros, and its charming architecture—nineteenth-century brick homes shaded by giant oak trees shrouded in moss. But over the years, the muddy banks of the Savannah River had gained the trappings of the maritime cargo industry: cranes, trucks, top loaders, an enormous rail yard. It had risen into the third-busiest container port in the United States, trailing only the combined operations of Los Angeles and Long Beach and the terminals of Newark.

The ports in Southern California and the greater New York area were notoriously short of room for expansion. Savannah, by contrast, was surrounded by open ground that could be turned into docks or warehouse space as needed, a central selling point to the ocean carriers.

The port already possessed nine berths for container ships and was in the midst of adding another—one big enough to accommodate the largest container ships—via a $600 million buildout. The plans included a new storage area with space for six thousand additional containers, plus an enlargement of the rail yard from five to eighteen tracks to allow more trains to pull in at once.

Yet even Savannah was so inundated by the crush of incoming containers that it was running out of places to put them.

On the day I visited in late September 2021, nearly eighty thousand containers were stacked along the shore in various configurations, like giant Legos strewn from the heavens. That was 50 percent more than normal, and dangerously close to the port's outer limit.

Many of the containers were stacked five high, which made every movement more complicated. With the docks full and the boxes piled into towers, locating the proper container for a truck waiting to haul it away typically entailed moving others around first.

The man in charge, Griff Lynch, executive director of the Georgia Ports Authority, was clearly frustrated by his powerlessness in the face of a situation whose constituent troubles extended well beyond his gates. With surrounding warehouses full and truck drivers scarce, many importers were simply leaving their boxes on the docks.

He took me to a part of his yard that was normally empty. On this day, it was jammed with seven hundred containers that had sat uncollected for more than a month.

"They're not coming to get their freight," Lynch groused. "We've never had the yard as full as this."

As he spoke, another giant vessel glided silently toward an open berth, the 1,207-foot-long *Yang Ming Witness,* loaded with more than ten thousand more containers. Lynch eyed it wearily, like a guy trying to empty his basement only to spot more boxes demanding space.

"Certainly," he said, "the stress level has never been higher."

Lynch had the no-nonsense demeanor of a man born and raised in Queens, New York. He had spent his professional life attending to the logistical complexities of moving cargo between land and water.

"I actually wanted to be a tugboat captain," he told me. "There was only one problem. I get seasick."

In his midfifties, he was suddenly contending with the most complex challenge of his career, a storm system whose intensity and contours were unparalleled.

The previous month, his yard had held 4,500 containers that had been there for at least three weeks.

"That's bordering on ridiculous," he said. "The supply chain is overwhelmed and inundated. It's not sustainable at this point. Everything is out of whack."

He showed me the construction site for the new berth—an army of backhoes shaping a giant sand pit. He stopped to show off the work underway at the railyard. He was seeking federal money to dredge the channel leading to the port, clearing the path for even larger vessels. Here was the fix seemingly emerging.

Then he gazed toward the water. A tugboat was escorting another ship to the dock—the *MSC AGADIR,* fresh from the Panama Canal—carrying more cargo that would have to be piled up somewhere.

"If there's no space out here," he said, indicating the stacks of containers, "it doesn't matter if I have fifty berths."

Among people who analyzed supply chain trends, some speculated that the pressure was about to relent. The shutdown of the Ningbo port had diminished the flow of ships headed to North America from Asia. A harrowing wave of COVID-19 had shut down Vietnam, halting factories that made clothes and shoes for the American market—another reason to expect a lull.

Lynch scoffed.

"Six or seven weeks later, the ships come in all at once," he said. "That doesn't help."

As conventional wisdom then had it, after a few more months, or maybe a year, everything would return to normal. The factories in China would resume their usual operations, allowing industrial plants and retailers worldwide to restock warehouses with products and components, curbing the impulse to stockpile. Americans would get back to offices and schools, reducing their demand for printers, desk chairs, and gaming consoles. All this would diminish the flow of containers bound for Savannah and every port. Pandemics ended eventually. So would this one, allowing truck drivers and dockworkers to return to their jobs.

But those who worked inside the supply chain were increasingly nursing suspicions that *normalcy* was a word that no longer applied. The confinement of humanity inside homes had accelerated the inevitable rise of e-commerce. People who had never before ordered something from Amazon had given it a whirl, gaining appreciation for the convenience. Older people, previously reticent to use apps to summon prescription drugs, had used COVID-19 as a reason to experiment. Even after the pandemic was gone, some of these habits would stick.

More e-commerce would add to the demands on the supply chain, requiring greater sophistication. The old days of pulling hundreds of containers off ships and trucking them to a single warehouse, for delivery to a handful of retail stores, would no longer cut it. Now, containers had to be unloaded at distribution centers that would deliver millions of custom orders to individual households—a far more complicated undertaking.

At the port, Lynch climbed an observation deck set in the center of the docks. He surveyed the sweep of shipping containers, the sun glinting off their multihued sides.

It was only September, and the Georgia heat was muggy and oppressive. Yet the Christmas season felt close at hand. Many of the boxes were surely full of wreaths, Christmas lights, and wrapping paper, along with countless presents.

Would all this stuff make it to stores and homes in time?

"That's the question everyone is asking," Lynch said. "I think that's a very tough question."

That was certainly Hagan Walker's question.

After ten days in floating purgatory, the *Maersk Emden* was finally permitted to enter the port of Long Beach.

Just after 1 p.m. on October 25, the vessel pulled alongside berth 134 at Total Terminals International, better known among dockworkers as TTI. The facility was controlled by MSC, Maersk's alliance partner.

Glo's order had completed the ocean portion of its journey, the 5,700 nautical miles that separated Ningbo from Long Beach.

But that left a continent still to be crossed, with myriad pitfalls to navigate along the way.

The next stage of the passage to Mississippi would depend on the orchestrated exertions of a group of people whose pay had long been a primary target in the cost-cutting designs of those running the supply chain: the dockworkers.

CHAPTER 11

"CRAZY AND DANGEROUS"

Life on the Docks

Before the container arrived, imposing order on the flow of cargo, longshore work was a feast-or-famine existence. On days when vessels pulled into ports, demand for labor was intense. The rest of the time, no one was needed. Given the imperative to maximize earnings whenever work presented itself, the docks were jammed with desperate longshoremen literally pleading for assignments.

"In America, it was known as shape-up," Marc Levinson wrote. "The Australians called it the pick-up. The British had a more descriptive name: the scramble. In most places, the process involved begging, flattery, and kickbacks to get a day's work."

In the English port of Liverpool, the price was often borrowing money at usurious rates from a crooked foreman with a side gig as a loan shark. The foreman had every interest in ensuring that those on the hook for loans received regular jobs, allowing him to collect what was owed via deductions from their pay.

Over decades, union mobilization combined with government intervention tamed the worst indignities of the morning scramble in many countries. The unions were focused on gaining job security for their members by taking over the functions of assigning work. They fought to increase the number of full-time positions while fending off demands from port managers to add casual workers—those who worked for lower wages on flexible schedules.

In the years after World War II, government boards in Britain and Australia assumed the process for assigning dockworkers. In the Netherlands, a series of crippling strikes prompted port operators to rely on full-time workers while largely disavowing casuals. And in the United States, a powerful labor movement turned longshore work into a steady way to make a living.

The rewards of that mobilization were evident as I surveyed the scene inside the cavernous union hall near the port of Long Beach just before six o'clock on a Monday morning in March 2022. The sun was still an unfulfilled promise on the watery horizon. Hundreds of longshore workers lined up in orderly rows, waiting to secure their assignments for the day.

The dockworkers craned their necks toward screens perched high on the walls, displaying the jobs on offer. There were positions for forklift drivers, for people who lashed containers together atop vessels, and for so-called swampers, who roamed around, attending to whatever needed doing—the busboys of the docks.

A bewildering chorus of announcers beseeched workers to step forward to a series of bank-teller-style windows at the far end of the hall to present themselves for the available assignments. One person was needed for "general dock work" at berth 126, starting at 6:30, and another at berth 136, to unload a container vessel freshly arrived from Ningbo. At berth 92, the *Dusseldorf Express*, a Bermuda-flagged container ship bearing cargo from Mexico and Central America, needed another longshore worker.

The people assembled there were the so-called Class A dockworkers, who had the right to labor on their own schedules. They had to work only once every thirty days, and for a total of at least eight hundred hours over the course of the year, to maintain guaranteed pension payouts along with health benefits and access to the best-paying jobs at the ports. They picked their assignments according to a process that gave priority to those who had worked the fewest hours that month. Most earned more than $100,000 a year, and many far more.

"You get in here, you're not going to want to go anywhere else," said Jesse Lopez, secretary and treasurer of the local longshore union. "You get a check every Friday, because there's work."

Lopez's father had been a longshoreman at the ports of Los Angeles and Long Beach. So had his father before him. The ranks of the union were full of such multigenerational stories.

Growing up in nearby Wilmington—a cluster of modest, Spanish-style homes on streets lined with palm trees—Lopez knew that mortgages and rent were covered largely by paychecks earned at the port, or in jobs that supported the enterprise. People repaired machinery, drove trucks, and prepared food for others engaged in moving cargo on and off ships.

The men—they were all men then—came home physically exhausted and often injured. They nursed ailing knees and strained backs, their fingers wrapped in bandages. They were drained by their perpetual proximity to danger; by the unshakable knowledge that any moment could bring a lethal accident. But they came home behind the wheels of their own cars, their money clips full of cash for groceries and toys and weekend outings.

Lopez's parents owned their house. He and his five brothers and sisters lacked for nothing of consequence.

"We had food on our table, clothes on our backs, and all our school supplies," Lopez told me.

The port was not only a source of sustenance. It was the centerpiece of the broader community. After their shifts, dockworkers stuck around to fish together off the docks, or drink beer in surrounding taverns. They looked out for one another, inquired about children and ailing relatives. They attended one another's weddings, christenings, retirement parties, and funerals.

When Lopez was thirteen, his dad taught him how to operate the controls of a crane. Still, he understood that he could not simply show up at the port to seek work. He had to be in the union.

One day in the early 1990s, his father clambered up a crane hovering over the docks and suffered a heart attack, plunging more than fifty feet. He tried to get back to work, but his body could no longer bear the strain. He died two years later. Lopez, then in his early twenties, assumed his father's union slot, under an arrangement that allowed the children of deceased workers to inherit their parents' rights to work.

He rose to the most exalted status among the union workforce: crane driver. He earned enough to move his own family—himself, his wife, and their two daughters—into a three-bedroom house with a swimming pool in a quiet community within reach of the port. They took vacations to New York and Cabo San Lucas.

"I'm able to take care of my family," he said.

He was grateful for his comforts. He was also clear on the foundation upon which they rested: a fiercely disciplined union.

The union was the key element that distinguished the dockworkers from every other laborer that had a hand in moving Glo's container from China to Mississippi. The longshore workers stood out as a group that had positioned itself to secure a reasonable share of the bounty. The container had been advanced as a tool for the executive class as it sought to move more cargo while diminishing pay for those engaged in the grueling and dangerous labor on the docks. The union had succeeded in ensuring that workers still got their piece of the action.

As Lopez wandered the hall, people came over to administer every conceivable form of greeting—fist bumps, pats on the back, high fives, handshakes, and hugs. As the secretary and treasurer of the local union chapter, he presided over the dispatch operation, and he seemed to know everyone.

He rubbed the shoulders of a man newly elevated from the ranks of the casuals after fifteen years in those trenches.

He embraced an older man whose face showed the scars of recent cancer surgery.

"I'm alive," the man said. "I'm blessed to be alive."

"You are, brother," Lopez replied.

They were standing near a bronze statue of Harry Bridges, the Australian-born labor organizer who had founded the ILWU back in 1937. His pugnacity on behalf of dockworkers had prompted the federal government to seek to deport him as a Communist.

Bridges had overseen a series of often-violent strikes in the years before the union's formal recognition. His favored tactic was the work stoppage. His people would halt loading and unloading, wounding the operations of the shipping companies in a demonstration of their indispensability.

The result of his campaign was a voluminous set of rules and conventions that limited which dockworkers could perform what tasks while denying the flexibility to move people around. Port operators and shipping companies chafed at the resulting regimentation, the obvious make-work provisions, and the restrictions on more efficient operations. But they assented because the alternative was the unrelenting threat of disruption. They ceded control as the price of peace.

The scene at the union hall in Long Beach was a direct outgrowth of the ferocious strikes waged by dockworkers along the West Coast almost ninety years earlier. In the decades since, the system of assigning jobs had been entrusted to some form of random drawing combined with seniority, under a process controlled by the union itself.

Lopez and his union peers understood that they possessed something that had become a rarity in the industrial American economy—jobs that provided a solid, middle-class existence. Even casual workers, who were guaranteed no shifts, still began at more than $32 an hour.

———

Among other laborers at the ports, the longshore workers tended to provoke animosity. They were often described as the aristocrats of the docks—an overpaid, entitled tribe that lorded it over everyone else.

"They treat us like we're nobodies," said Anthony Chilton, a truck driver who hauled containers between the Southern California ports and warehouses to the east.

Another truck driver, Marshawn Jackson, said the dockworkers acted as if the ports were their exclusive preserve.

"They're the rudest people in the world," he told me. Like 'Don't say nothing to me.' Like, 'This is ours. You're just here.'"

These sorts of depictions were reverberating in early 2022, as the ILWU's contract came up for renewal, provoking fears of a labor impasse that would deal yet another blow to the supply chain.

The union represented twenty-two thousand dockworkers along the West Coast. Nearly three-fourths were employed at the ports of Los Angeles and Long Beach. Even in normal times, they possessed enormous leverage, given their crucial role in moving cargo in and out of the world's largest economy. In the midst of an epochal supply chain disruption, the longshore workers wielded greater power than ever. Yet their bargaining position was derived directly from their capacity to threaten the livelihoods of everyone else who depended on the ports. Their strength and the vulnerability of other workers were two sides of the same coin.

This explained the resentment from truck drivers. Their ability to earn a living by hauling containers rested on the willingness of the dockworkers to load and unload the boxes.

"Every time there's a contract up, things slow down," Chilton complained. "We always blame the longshoremen. They slack off, take breaks, call in sick."

As a rule, the executives of shipping companies and logistics businesses were too fearful of angering the union to discuss labor issues on the record. Privately, they described dockworkers like trolls guarding a bridge.

To Lopez and his fellow longshore workers, such portrayals were a source of grievous indignation. The accusations ignored a central feature of their lot—the grueling nature of their jobs, and the ever-present possibility that any day could end in catastrophe.

"You don't get hurt down here," Lopez said. "You get killed."

Twice, he had seen longshore workers crushed to death by falling containers. More times than he could recall, he had to knock on the door of a union peer and deliver word to their spouse and their children that "their loved one is not coming home tonight."

This part of the story went missing from cartoonish characterizations of overpaid dockworkers. And this was a crucial detail unknown to consumers as they waited for their goods to arrive without having to contemplate the vulnerable human beings who kept commerce moving.

"It's crazy and dangerous," Lopez said. "Anything could happen. I've seen lights fall off cranes and land right next to people. Even if you're wearing a hard hat, that's not going to save you. You're physically tired, but also mentally tired. You've just got to work down there with your head on a swivel."

The pandemic had added a new layer of angst. The people working the docks at American ports were deemed essential workers, meaning their labors were required to keep the national economy moving. Among the ILWU ranks, at least two dozen members had died of COVID-19 by early 2022. Even so, the longshore workers continued to move record volumes of cargo.

"When everybody else was shutting down," Lopez said, "we didn't stop."

There was something perverse about a conspicuously successful union having to defend its middle-class wages against a backdrop of American despair.

Over the previous seven decades, labor unions had been decimated by American employers through a barrage of lobbying, court decisions, rulemaking, and strong-arm tactics. They had been bypassed by hiring practices that emphasized part-time and temporary workers over full-timers—the sort of workforce that provided the flexibility celebrated by McKinsey in preaching the gospel of Just in Time. And the unions themselves had failed to organize workers in emerging industries like technology.

The result was an economy in which only 10 percent of American workers were represented by a union. That compared to 20 percent in 1983. And it was down from a peak of 28 percent in 1954.

That trend represented the triumph of corporate executives in their mission to dilute the power of labor unions. They waged that battle not for vague ideological reasons, but out of a pragmatic interest in paying their workers less. The economic literature showed that union laborers earned between 10 and 30 percent more than their nonunion counterparts. The spread was especially wide among lesser educated, lesser skilled workers.

Decades of data also brought home how slashing paychecks for ordinary workers was an excellent way for executives to boost their own compensation. Companies whose workers were represented by unions were far more likely to limit pay for top executives as a way to avoid angering the rank and file ahead of contract negotiations. Companies that lacked union representation, on the other hand, tended to hand executives extravagant raises and bonuses even when they failed to boost the profitability of their businesses.

As a result, union power had declined over the same decades in which the chief executives of publicly traded companies had seen their pay skyrocket. In the mid-1960s, when union membership was near its peak, the average chief executive had earned about twenty times as much as the typical worker. By 2021, CEOs were netting nearly four hundred times as much as rank-and-file workers.

Workers had long wielded collective action as their primary means of seeking their share of the gains of capitalism. Once union power was eviscerated, employers were able to divert more of the profits to themselves.

Between 1948 and 1979, American workers roughly doubled their productivity, a measure of how much economic value they generated through an hour of labor. Over the same period, their pay also doubled. What was good for companies proved good for their employees.

But over the following two decades, that relationship broke down. Productivity continued to increase, expanding by another 62 percent between 1979 and 2020. But worker compensation increased by less than 18 percent.

Freed from the nuisance of unions, the executive class pressed down pay while helping themselves to the money that had previously gone to workers.

Beyond the social implications of this widening inequality—the scarcity assailing communities, the breakdown of faith in democratic institutions—taxpayers were footing the bill for the obliteration of labor power. You could see it in the ranks of those who depended on food stamps to feed their families. More than 9 million American families relied on the program at some point in 2018. More than three-fourths of those families included at least one person who had a job.

A job was no longer reliable insurance against poverty.

The term of art to describe people caught in that circumstance was "the working poor." Their ranks included people employed at warehouses run by companies like Amazon. There, an estimated thirty-eight thousand workers relied on some form of government benefits, from food stamps to Medicaid, according to an analysis conducted in 2021. The previous year, Amazon had logged $21 billion in profits, sending its shares soaring, and making its founder, Jeff Bezos, the richest person on earth.

In essence, taxpayers were subsidizing the winnings harvested by executives and shareholders through the dismantling of labor unions. And consumers were availing themselves of goods that were made cheaper by the desperation of the people moving them around.

Part of the calculation that drove Walker to rely on factories in China was the implicit understanding that transportation from Southern California to Mississippi was an almost trifling expense. That reality rested on the fact that most of the people laboring along the way were not represented by a union. They brought little leverage to the pursuit of compensation.

The dockworkers represented an outlier in the story of downward mobility. They did not need handouts from anyone. Yet in the zero-sum game of the American economy—in a supply chain hollowed out by reverence for Just in Time, and optimized for the betterment of shareholders—their gains were routinely cast as a threat to everyone else's prosperity.

In the conversation about their looming contract negotiations in the summer of 2022, the dockworkers were typically presented as the latest wild card bearing down on a supply chain struggling to return to normal.

"Delays in reaching an agreement, or an outright stalemate, could exacerbate our current inflation challenges, return us to the stock-outs of the early COVID era, and bring the global economy to a standstill," warned John Drake, an executive at the US Chamber of Commerce, the lobbying organization for large businesses.

But the very concept of normalcy had been downgraded along with the wages and working conditions of millions of people whose exertions were critical to the movement of cargo. Normalcy was workers not earning enough to secure health care or buy groceries. Normalcy was millions so desperate for paychecks that they would accept danger, indignity, scarcity, and family stress as part and parcel of their jobs.

The dockworkers had managed to avoid this fate through disciplined organization. They had secured compensation that was above normal, with terms generous enough to banish fears of household duress. Yet in the everyday narrative peddled by business groups, this made them gluttonous. *They* were the reason for product shortages and inflation, and not the ocean carriers then logging record profits.

"Previous labor disputes at ports cost the U.S. economy upwards of $1 billion to $2 billion per day," declared forty-nine trade associations representing major importers in a letter to the White House urging the Biden administration to broker a deal. "To say the stakes are even higher today is an extreme understatement, as even a short slowdown or shutdown will disrupt already fragile supply chains and compound inflationary pressure."

The supply chain was indeed fragile. And dockworkers were positioned to throw a wrench in the gears. Same as ever, this was their most potent tactic. The implications for the broader economy were potentially grave.

But the longshore workers were not the villains in the story of supply chain dysfunction. They were the survivors. Lopez and the rest of his union were the descendants of those who had endured a decades-long campaign by the shipping industry to reduce them to insignificance.

The container had been a major advance in this battle, delivering a monumental leap in efficiency and sharply reducing the need for longshore workers. But that was not the end of the story.

Robots were poised to finish the job.

Three years after Malcom McLean's first voyage out of Newark, the costs to ship many goods remained so high that they reached one-fourth the expense of manufacturing the product. As much as three-fourths of the costs of ocean shipping were captured by activities undertaken while ships were docked.

The container shrunk those expenses by limiting the time that a ship needed to remain in port. By standardizing the loading and unloading process while allowing for greater use of machinery, the container diminished the need for human hands.

Harry Bridges, the founder of the ILWU, understood the pressures keenly. He also grasped that containers were an unstoppable force. Rather than engage in a futile effort to halt technological progress, he focused on what he could extract for his members.

In the early 1960s, Bridges championed a controversial deal that permitted the installation of cranes and the other equipment needed to load and unload boxes. In exchange, he gained robust wage guar-antees for those fortunate enough to remain employed.

More than half a century later, dockworkers around the globe were confronting the latest innovation engineered to marginalize their importance to the running of ports—robots and other forms of automation.

In 2016, I visited a Maersk container shipping terminal in Rot-terdam, Europe's largest port. An enormous container vessel, the Danish-flagged *Mette Maersk,* was tethered to the dock. I watched robotic arms grip containers, lift them off the ship's deck, and de-posit them onto stacks below with thunderous booms. A fleet of self-driving trucks glided quietly through the yard.

John Arkenbout had lived through the transition. When he began his career as a longshoreman at the port a quarter century earlier, he had worked outside, enduring the unrelenting wind and drizzle of the North Sea. He had used his body to lift huge bricks from a pile and drop them into rope sacks that a crane operator lifted skyward, right over his head.

That scene had become a memory. Arkenbout, then in his early fifties, and most of his fellow longshore workers mostly hewed to the comfortable, climate-controlled offices nearby, staring at a screen and using a joystick to manipulate the cranes and other machinery. You could not compare these two versions of working reality and yearn for the past.

"Before, it was physically taxing," Arkenbout told me. "Now, it's more mental."

But it was also far less secure.

The ranks of his union, FNV Havens, had dwindled from about 25,000 members in the 1980s to some 7,000. And with more robots on the way, union leaders were girding for another 800 job losses over the next several years.

I was visiting Rotterdam to explore how the shifting of produc-tion to low-wage countries like China had stoked anxiety and anger within working-class communities. Robots at first struck me as a wholly different subject. But the longer I spent with the dockworkers in Rotterdam, the more I came to see that automation was central to understanding the alienation and anger seething among people who earned their living with their hands.

Over the years, Arkenbout and his fellow dockworkers had seen the Rotterdam port fill with truck drivers from Eastern Europe, who worked for a fraction of the wages commanded by their Dutch counterparts. More recently, they witnessed the logical extension of this trend—trucks with no drivers at all. They had watched the ports hand the work of cleaning out containers to North African immigrants.

They were cognizant that the ultimate winners in these developments were not the low-wage employees who were getting the work, but the conglomerates that dominated the shipping trade.

Trade liberalization, immigration, containerization, automation. They were all instruments wielded in the service of a consistent objective—stiffing workers and transferring the proceeds to the multinational companies that dominated the supply chain.

The work of moving containers between ships and docks was more important than ever. At least for the time being, the dockworkers' union still derived leverage from threatening to withhold their labor, adding to the worries of shippers like Walker that their orders would suffer delays.

But the human beings who had for generations performed this work were increasingly incidental to the designs of the companies running the ports. Robots were the ultimate insurance against Harry Bridges's old ploy, the work stoppage. Robots did not get stuck in quarantine. Robots did not demand better pay.

"I'm an employee," Arkenbout told me. "In the end, I'm nothing. When they don't need me anymore, I'm nothing."

On the West Coast of the United States, the contract talks were certain to come down to bargaining over the pace of automation.

Lopez and his fellow longshore workers had little to say on this subject. The union was tight-lipped, unwilling to tip its hand ahead of talks. But the same formulation that Bridges had applied to the container seemed certain to dictate the union's strategy. It would focus on the compensation it could gain for its members as the price of its inevitable agreement to allow more robots on the docks.

They could not keep automation at bay forever, this the union understood. Technology would win, because it always had. It made cargo handling more efficient, which advanced Just in Time, which was good for shareholders.

For the dockworkers, the goal was simply slowing down the future for as long as possible.

CHAPTER 12

"IS IT WORTH EVEN GETTING UP IN THE MORNING?"

The Unremitting Misery of the Dray

Because the longshore union was in control of the system used to deploy workers in Southern California, Hagen Walker could take for granted that, once his container finally reached land, someone would be available to unload it from the ship.

After the *Maersk Emden* pulled up to the TTI terminal, a crane operator manipulated the controls to lift Glo's box off the deck and deposit it into the stacks below. There the box sat for four days, awaiting customs clearance.

Then came the next phase of the journey, a truck ride to a nearby warehouse, where Glo's container would be unloaded and repacked into a trailer for its ride across the country. On the paperwork prepared by Freightos, the platform Walker used to book his container, this movement was so inconsequential that it did not even merit a line item. In reality, it required the labor of one of the most volatile realms in the supply chain.

A truck driver would need to get through the gates and haul the container away.

———

Trucking, in contrast to the longshoremen's work, was a free-for-all industry with no central authority in charge of matching supply and demand. Once Walker's box landed on the dock, there was no telling whether a truck driver could be quickly engaged to haul it to the warehouse where it would be unpacked, or whether that might take days, or even weeks.

Sometimes a massive glut of drivers was eager for work. Other times, drivers were few while boxes were many.

Here was the volatile situation navigated by drivers like Marshawn Jackson. He did not have a union behind him. What he had on his best days was freedom. He owned his own truck, selected the loads that suited him, and took orders from no one. He worked as a dray operator—trucking lingo for a driver who pilots their rig over relatively short distances, hauling containers between one point and another. He set his own schedule and determined his own routes as he navigated the tangle of freeways linking the ports of Long Beach and Los Angeles to the low-lying warehouses to the east, in the flatlands of the Inland Empire.

But on his worst days, which were many, Jackson had only his status as an independent contractor. Which is to say he had nothing.

He had no guarantees of work, even as the bills demanded payment: the insurance for his rig, the fee to park it at a yard near his home, and the unrelenting expenses of everyday life, from his mortgage to his phone bill to the costs of keeping his family fed.

Even on days when there were plenty of jobs and at soaring rates, he still contended with unceasing agita. During much of the pandemic, the traffic at the ports was so awful that dray operators were forced to sit for hours at every conceivable point along their journey.

They waited at the gates before they could gain entry.

They waited for hours, stuck inside their cabs—and with no access to restrooms guarded vigilantly by dockworkers—before they could pick up their boxes.

They waited for clearance to exit, and they waited outside warehouses before they were able to enter yards and drop their boxes.

Some days, they waited for hours before they could pick up a chassis, the wheeled trailers on which the containers rested as they pulled them behind their tractors.

Other days, they got stuck waiting for safety inspections, or for mandated repairs, or for the chance to swap out some failing piece of equipment before they were permitted to leave.

They waited in an information vacuum, not knowing why the ships were stuck, or why the containers took so long to load, or why some trucks had to sit there all day while others were waved through.

"They don't tell us what the holdup is," Jackson said. "It's so up and down. It's confusing. Nothing's clear. All I know is 'Pick up at point A, and drop off at point B.'"

The basic infrastructure of the ports could not manage the surge of containers arriving from Asia. The dray drivers were the ones who suffered the consequences most acutely.

Jackson got used to pulling out his neck pillow and sacking out behind the wheel while his rig sat on the pavement. He slept fitfully, gnawing on the injustice of his situation. He and most of the other dray drivers got paid per load, and not for their time. The longer they waited, the less they were making per hour.

Here was the dirty secret of a supply chain long governed by the ruthless efficiencies of Just in Time. It treated the people who did most of the work like their own time was at once limitless and worthless.

They got stuck waiting, their time unpaid, staring at a line of other vehicles in front of them, and wondering when the agony would relent.

This was why thousands of dray operators were quitting in the fall of 2021, just as Walker's container landed on the docks in Long Beach needing a ride.

In Southern California, drivers accustomed to hauling three

containers in the course of their day were lucky to complete one job, as the traffic jams consumed their hours.

"How do you convince truckers to work when their pay isn't guaranteed, even to the point where they lose money?" asked one veteran driver, Ryan Johnson, in a social media post that went viral. "Nobody in the supply chain wants to pay to solve the problem."

In the center of the country, the rail system was overwhelmed, yielding traffic jams that were ensnaring dray operators for hours. Their take-home pay had plunged by 20 percent even as rates for loads were soaring. From Chicago to Dallas, one-fourth of dray operators were quitting.

On the East Coast, at the port in Charleston, South Carolina, incoming containers were increasing by one-fifth, yet local dray drivers were picking up nearly one-tenth fewer boxes as congestion slowed the works.

And down in Savannah, Griff Lynch, the man in charge of the port, stared at the stacks of uncollected containers piling up on his docks and pinned much of the blame on a shortage of drivers.

"This has been building for a long time," he told me. "What parent is encouraging their child to become a truck driver?"

For Jackson, driving a truck was a way out of the challenging circumstances that framed his life.

He was an African American man raised in South Central Los Angeles, which had developed as a place of last resort for Black and Latino families excluded from other parts of the city by decades of racist housing covenants and discriminatory mortgage policies.

Only one-fourth of households owned their own homes. More than two-fifths were officially poor. Parks and shops were sparse. Gang violence was inescapable.

"You get used to seeing things," Jackson told me. "All you can do is pray you can make it out."

His mother worked at the post office, a source of solid paychecks in a neighborhood dogged by chronic shortages of decent jobs.

He helped his grandmother with her family hair care products business. He packed boxes in a warehouse when he was only ten. After high school he joined the company full-time and managed orders. He gained exposure to the business side, accompanying his uncle to trade shows where they looked for new products. He even went along on a reconnaissance trip to Shanghai.

Jackson had a head for numbers. He would have stayed on at his grandmother's company long term. But then came the terrible downturn that followed the financial crisis of 2008. Their sales evaporated, and they eventually shut down the company. Jackson was suddenly in need of a reliable way to support himself, his partner, and their then-infant daughter. A friend told him there were good jobs to be found in long-haul trucking.

As a child, Jackson had been fascinated by trucks. He liked the thought of seeing life beyond Los Angeles. And he was especially enticed by the handsome compensation promised by Swift, the national trucking firm that was permanently recruiting new drivers. The company beckoned with a pitch promising liberation—from financial worries, from the mundane existence of waking up in the same place every day, from the limited opportunities found in communities like South Central.

Its slick advertisements cast driving a truck as a noble profession central to the American economy.

"Without trucks, this country would stop," said one Swift driver in a promotional video broadcast in 2010.

"It can be fun, it can be exciting, and it can be profitable," said another company driver as the video displayed trucks winding a curve through rugged mountains. "Before, I was struggling, worrying about paying my bills. I don't have no worries, man, no worries. My kids are happy, my wife is happy, I'm happy."

Jackson bought in.

Swift fronted him financing to cover the $6,500 cost of a three-week training program needed to gain a commercial driver's license. In exchange, he committed to driving for the company long enough to pay back his tuition via deductions from his pay.

The training program was held in Phoenix. He rode a Greyhound bus for eight hours across the desert and crammed into a grubby hotel room plagued by scorpions with two other fledgling drivers. They practiced on beaten-down trucks that lacked air-conditioning despite summer heat exceeding 115 degrees.

Soon, he was earning $1,000 a week hauling trailers full of products from a dollar store distribution center in San Bernardino, California, to warehouses in Phoenix and back. But paying back his tuition lopped off nearly a third of his pay. He later learned that he could have trained at a private school that would have charged him only $2,000. By then, he was irretrievably on the hook to Swift.

"Once you get over there, you feel like you're stuck," Jackson said.

This was a common realization in an industry dependent upon unceasing flows of new recruits to replace drivers who quit. Throughout the American trucking industry, the average turnover was nearly 100 percent, meaning that the typical firm had to replace its entire workforce over the course of a year.

That number reflected many factors—the grueling nature of the work, the difficulties of balancing family responsibilities, and disillusionment as visions of open road adventures succumbed to the grim monotony of warehouses, truck stops, and portable toilets.

But the high turnover also attested to the predatory financing that was central to the trucking game, from the training program rip-off that Jackson suffered to commonplace arrangements in which trucking firms leased and sold rigs to drivers at rates of interest that could make a loan shark blush.

"Many of the workers who try out trucking every year do so under a modern form of debt peonage," wrote Steve Viscelli, a labor expert at the University of Pennsylvania, and himself a former truck driver.

Jackson soon discovered that his yearning for freedom was tempered by the reality that he answered to dispatchers who treated him like a cog in a giant delivery machine.

He picked up refrigerated trailers full of freshly harvested lettuce in central California and hauled them across country to a Dole distribution center in North Carolina. Then he frequently ventured up the East Coast to New England carrying other freight, and back across the continent to warehouses in the Pacific Northwest, on loops that lasted two and three weeks.

His weight swelled beyond three hundred pounds as he subsisted on fast food and rarely exercised. He had no time to stop and enjoy the scenery. He collected only fragments of experience gleaned from a life dominated by the imperative to keep moving.

"Sleeping and driving," he said. "That's all I really did."

He remembered the white woman at a truck stop in Iowa who screamed at him, demanding that he leave rather than park in her lot while he took a half-hour break. "She called me every kind of racial slur," he said.

He remembered the kindly, gray-haired driver who led him across North Dakota in a frigid blizzard, through dangerous whiteout conditions, telling him to keep rolling slowly, never stop, lest the subzero temperatures turn his fuel into jelly.

He was by then making $1,600 a week. But he was handing half of that back to the company to cover the payments on the truck he had begun leasing, enticed by the pitch that he would be an owner-operator. He later learned that he was paying an interest rate of 29 percent. When his required repairs, Swift demanded that he take it to one of its shops, where he paid four times as much as he might have at an independent establishment.

Mostly, he remembered the strains of being away from his family. He would return home exhausted and needing a break, only to discover that his wife was equally spent, having taken care of their young daughter solo.

In 2016, when his daughter, Bailey, was graduating from kindergarten, he implored the dispatchers to schedule him to be at home, just for that day, so he could attend a ceremony. He gave the company three weeks' notice.

But one dispatcher, a gruff former Marine, mocked him. "This is what you signed up for," he said.

On the day of the ceremony, Jackson was out on the road.

"I felt like I was letting my family down," he said. "It bothered me. It changed my whole outlook."

He drove back to Southern California and handed over the keys to his rig, accepting a colossal financial loss in exchange for his escape from Swift. He had already made payments exceeding $50,000, but his outstanding balance remained in excess of $100,000—this, on a vehicle that had more than six hundred thousand miles on the odometer, and that could be purchased new for about $90,000.

"It wasn't a good feeling," he said, "but I was happy to be done with it."

He bought his own truck, a used model, for $30,000 and picked up routes for himself from brokers. He was still driving long-haul, but he went no further than Texas, limiting his time away to three days.

And then he figured out how to sleep at home every night. He began taking dray assignments, working in and out of the ports.

He picked up jobs using an online platform called the Dray Alliance, a venture funded technology company that paired up drivers with shippers.

He and his family moved into a rented apartment in San Bernardino. Then they bought a modest house just off the freeway.

He traded in his old truck for a newer model. Then, he bought two more, hiring drivers to handle the extra loads. He was running a small business, cashing in on the seemingly unlimited work for dray operators during the first two years of the pandemic, even as he spent too much time sitting in lines at the ports.

In the summer of 2022, when he was thirty-seven, he signed off on purchasing a four-bedroom, newly constructed home with space for a planned swimming pool in Jurupa Valley, a growing community carved into the desert in Riverside County.

The house was a twenty-minute ride from the yard where he parked his truck and a world away from South Central Los Angeles.

Just before four o'clock on a Tuesday morning in mid-September 2022, the sky still black save for the reddish glow of the freeway, Jackson rolled over in his bed at his home and reached for his iPhone.

He clicked on the Dray Alliance app to try to arrange loads for himself and his two drivers.

Only a couple of months earlier, he could take his pick of dozens of possible loads paying premium prices. On this morning, he absorbed a message that had become familiar and disheartening.

"No jobs available."

What jobs did show up were far less gratifying than before. In late 2021, dray operators were earning about $700 to haul a container from San Bernardino to the Port of Los Angeles—a seventy-mile journey. On this morning, that same run brought only $500, even though the price of fuel had increased.

Yet even at those sharply reduced rates, every job was gone in an instant, snapped up by a desperate driver with bills to pay.

"They know we've got to keep working," Mr. Jackson said. "That's how they take advantage. We've got to survive."

His morning and afternoon were already booked with assignments he had secured the day before. But the rest of the week was open. Over the course of the day, he refreshed the app constantly, desperate to secure more work.

He refreshed after he pulled his tractor trailer into a nearby storage yard to pick up an empty shipping container, and again while he rolled down the freeway, en route to the Port of Los Angeles—one hand on the wheel, one hand on his phone.

He refreshed in the yard where he dropped the empty box, and a dozen more times while he waited for a crane to deposit another container on the chassis behind his rig, this one loaded with Mattel toys freshly arrived from factories in Asia.

He refreshed while he fueled his rig, and while he walked to the restroom.

Each time, the same result: no jobs.

"You reach a point where you're like, man, am I even making money?" Jackson said. "Is it worth even getting up in the morning? I'm sitting here on the 57 freeway, looking at my phone, and I'm like, am I gonna be in an accident? But I've got to hit refresh, refresh, refresh."

The sudden disappearance of work reflected the fact that fewer containers were arriving at Southern California's ports.

In part, this was because American demand for goods was finally waning, as inflation limited spending power. It also reflected how major retailers were bypassing Los Angeles and Long Beach, instead shipping to East Coast destinations like Savannah to circumvent potential slowdowns as Lopez and his fellow ILWU dockworkers faced off with port operators over the terms of their new contract.

At press conferences, Gene Seroka, who ran the Port of Los Angeles, celebrated the taming of the traffic jams, the end of ships stuck floating off the coast.

But the result was more volatility for dray operators, more worries about bills, more reason to scrimp.

Only six months earlier, Jackson had been bringing home as much as $1,800 a day, before accounting for overhead costs like fuel and maintenance. On this day, he would be fortunate to make $1,000.

As he navigated five lanes of traffic on the way to the port, he donned headphones for most of the journey, speaking by phone with his wife as they fretted over whether they would still be able to close on their new home, given the sharp fluctuations in his income.

He worried about whether he could line up enough work to sustain the two drivers he employed, and whether they would bring in enough to cover the expenses of operating his pair of other rigs.

He wondered when he would ever be able to take his wife and daughter, now thirteen, on another vacation.

He contemplated the tenuous nature of American upward mobility, the forces that seemed to be pulling him backward.

"The way we're living is hard times right now," he said. "You've still got to smile through it. You've still got to be positive. But, man, I'm dealing with a lot right now."

Though the Inland Empire lay roughly sixty miles east of the ports of Long Beach and Los Angeles, its clusters of warehouses amounted to an extension of the docks. Here, major retailers stashed the products delivered from Asia on container ships—clothing, shoes, furniture, electronics. They used local distribution centers as jumping-off points to supply consumers across much of the American West.

In the same way that massive slaughterhouses turned Chicago and Kansas City into crucial hubs for rail cargo in the late nineteenth century, the Inland Empire had burgeoned into a dominant center of distribution in the age of big-box retail and e-commerce.

At 5:43 a.m., the sun still a vague suggestion to the east, Jackson sat behind the wheel of his enormous blue Kenworth tractor in the city of Ontario, a smattering of shopping malls, suburban subdivisions, and warehouses. He guided his truck into a Shell station across the street from a Jack in the Box restaurant.

Diesel was selling for $6.19 a gallon, an eye-popping number. He put $100 in the tank, enough to get to Los Angeles to drop off the empty trailer he had picked up earlier that morning from a warehouse for LG, the home appliance company.

Fifteen minutes later, he was headed west on Route 60, the traffic building. He braced for what the day would bring. Even with compensation way down and the traffic jams a memory, the delays and glitches that had plagued dray driving for years remained.

Three times in the previous week, Jackson wound up on so-called dry runs, meaning a journey that had to be aborted because of some kind of mishap—paperwork not in order, a pickup appointment made incorrectly, a shortage of some key piece of equipment. For his trouble, he headed home with a $100 fee that barely covered the cost of gas.

During the worst days of the pandemic, Dray Alliance's algorithm had proved excellent at anticipating congestion, steering drivers to jobs where traffic was not bad, Jackson said. But its data management system could do nothing to redress a basic lack of work.

At twenty minutes past seven, the sun now vivid and gathering force, Jackson pulled into the container storage yard near the port, rumbling over bumpy pavement. The yard looked out over an oil refinery. He backed into a space between two other containers, stepped out of the cab, and turned a crank handle to lower the landing gear of the chassis. Then he detached it and left it there.

He quickly found the empty container that he was picking up. But he noticed that the chassis that held it was painted a fading tan—an indication that it was old. That could be a problem, triggering an inspection.

He drove to the port, entering the gates of APM Terminals—part of the Maersk empire.

As he waited for the security guard to check his credentials, he refreshed the Dray Alliance app. No jobs.

The security guard waved him through. A machine spat out two paper tickets—one to hand to the dockworkers overseeing the collection of his empty, and another for the loaded box he was picking up.

A few minutes later a dockworker driving a top loader—a machine that lifts containers—motioned Jackson forward into an appointed space so he could pluck the box off the chassis and add it to a stack.

Jackson scanned his phone for his next destination: space E162. The name was painted in white on the dock. He pulled in tight, his passenger-side mirror grazing the container parked to his right. A crane deposited his box onto his chassis. It landed with a boom and then the groaning of grinding steel.

There were no trucks in front of him, none behind him. He had visions of getting the container up to the Mattel distribution center in San Bernardino in time to grab a meal, his first of the day, before heading back to the port for his second run.

Then, just before he reached the exit, a port worker noticed the old chassis and diverted him for a spot inspection.

A mechanic barked at him to stop, and then approached the cabin.

"Just giving you the heads-up that your connections are all fucked up," he said.

Jackson sighed. "This right here, could basically slow up the whole day."

Of the fifty-eight thousand chassis maintained by a collection of equipment companies at the ports of Los Angeles and Long Beach, more than two thousand were in some state of disrepair. The mechanics with the skills needed to fix them were in chronically short supply.

Jackson sat in his cabin, leaving the engine running to provide air-conditioning against the mounting heat.

"Hey!" the mechanic shouted. "Cut off the truck!"

Jackson complied. For an instant, he allowed himself to gaze through his windshield at the water, serene and blue. Then he pulled out his phone and tried to refresh the app. But this part of the port had very weak cell coverage. He could not get online, a cause for alarm. If any jobs came on the board, he would miss them.

For more than an hour, the mechanic wielded a buzzing drill as he poked at a twisted hunk of metal flapping loose from the front end of the chassis. He called over another mechanic. The two men stared quizzically at the undercarriage.

Jackson grew agitated. The longer the delay, the greater the odds that they would pronounce his chassis unfit for the road and send him back to the "flip line," where he would have to wait for another chassis, potentially for hours.

Though chassis were stacked one atop another at a storage area, most were spoken for by the largest importers like Amazon.

On this day, the port gods smiled upon him. The mechanic affixed a sticker to the front of the chassis indicating that it was highway ready. Jackson proceeded toward the terminal exit.

Once back in cell range, he refreshed the Dray Alliance app and was horrified to see that he had indeed missed a job while he was beyond connectivity.

He pulled up to a truck stop in Long Beach, added another $400 in diesel to the tank, and headed to the restroom for the first time since he got on the highway before dawn.

By eleven, he was back on the freeway, returning to the Inland Empire to drop off the container full of toys for Mattel. He used one hand to shovel popcorn into his mouth. Then he put the bag on his console, picked up his iPhone, and refreshed. No jobs.

He rolled past a warehouse for Lean Supply Solutions, a billboard for Fastevict.com ("representing landlords only"), past a tent city full of homeless people clustered under a freeway overpass, past super-stores and drive-throughs and check-cashing outlets.

Many truck drivers obsessively caffeinate, perpetually afraid—and not without reason—that their hours behind the wheel will lead them into a state known as highway hypnosis. But Jackson does not drink caffeine. "I drink a lot of this," he said, taking a swig from a bottle of Fiji water.

To keep awake, he relies on the vibrations of his sound system.

"It gets loud in here," he said, cranking up the dial on the Isley Brothers classic "Work to Do":

I'm taking care of business
Woman can't you see?
I've gotta make it for you
And I gotta make it for me.

He cut the music as his phone vibrated to life. One of his two drivers was calling to report that he had accepted a job from Dray Alliance to drop off an empty container at the port and was headed back to the Inland Empire. He was bobtailing it—trucker parlance for pulling nothing behind his tractor.

Jackson was calm and unflappable, but this information clearly upset him. He had arranged for that driver to pick up a load at the port that evening. He should have waited to do both jobs at the same time, on one round-trip journey. Instead, he was burning fuel on two separate round trips, at Jackson's expense.

"How does that cover the cost of me paying you?" Jackson demanded. "The rates are down. It's slow, bro."

Heavy clouds hung low over the San Bernardino Mountains as he arrived at the Mattel distribution center just after noon. He dropped the container, picked up an empty, and was back on the freeway for the return leg to the Port of Los Angeles.

Just outside Whittier, on the 695, traffic ground to a stop. Jackson groaned. But then the flow resumed, and he arrived at the port with time to spare for his first meal of the day.

He rolled slowly through the cracked streets of Long Beach, looking for a stretch of empty curb long enough to park a tractor trailer. He found one around the corner from a donut shop, next to a parking lot full of broken-down chassis.

He used his phone to place an order via Uber Eats: a Chipotle bowl with chicken, brown rice, and avocado—an alternative to the greasy, unhealthy fare that prevailed at truck stops. He walked to the corner and waited for the delivery guy on a scooter—the supply chain for the people behind the supply chain.

By four, he was back at APM Terminals. He rolled down the window to show his credentials to the security guard, and the smell of the sea wafted into his cab.

"Sometimes, being on the road all the time, the sunshine and the breeze makes you feel better," he said. "It's kind of what keeps people sane."

He drove back into Long Beach and parked at the same spot, retiring to the bunk in the back of his cab. No use sitting in rush hour traffic. Better to stay here and rest. He watched a movie on his iPad but quickly fell asleep.

By the time he emerged just after six, the streetlights twinkled in the golden dusk.

Back on the freeway, he called his wife, his voice low and quiet as they exchanged worries about whether they would be able to close on their new home. The underwriter was demanding extra documents, concerned about the blurred boundaries between his business, now struggling, and their personal finances.

Darkness filled his cab as brake lights flickered ahead. A blue Tesla changed lanes aggressively, darting in front of him, forcing him to brake sharply. He and his wife were struggling to make out what lay ahead.

"People are like, 'If you get through this point, you'll be okay,'" Jackson said. "And I'm like, 'How long is this point going to last?'"

Hagan Walker's container did not have to wait long for a dray.

Four days after it landed on the dock at Long Beach, a driver working for a company called Pudong Prime International Logistics arrived at the terminal and hauled the box to a warehouse in Costa Mesa, near the John Wayne Airport. There, despite Walker's fears of the consequences of floor loading, a crew of workers quickly unloaded the container and repacked Glo's order, depositing the cartons onto pallets and into a fifty-three-foot trailer that could be pulled by a long-haul truck.

And then the trailer sat for two days, with no drivers available and a continent still needing to be traversed.

CHAPTER 13

"BUILDING RAILROADS FROM NOWHERE TO NOWHERE AT PUBLIC EXPENSE"

How Investors Looted the Locomotive

The journey Hagan Walker booked from China included a rail segment extending from the Port of Long Beach to Memphis, placing his container within striking distance of his warehouse in Mississippi.

In a rational world, the railroad was the best way to move a container across the country.

This was clear on environmental grounds alone. As alarm grew over the intensifying threats from climate change, railroads were a central part of the solution. Mile for mile, shipping freight by train spewed only one-tenth of the greenhouse gas emissions yielded by trucks.

As an economic proposition, rail was the better option, too. Moving freight by truck cost more than three times as much as transporting it by train. And railroads appeared an even better value once broader societal interests were factored into the equation. Trucking entailed greater wear and tear on bridges, traffic jams that wasted hours, and highway accidents that cost lives. These sorts of costs were eight times greater for trucking than for rail.

Yet in the world as shaped by shareholder interests, rail was a realm plagued by operational perils, labor shortages, malfunctioning equipment, and delays.

American railroads had been looted from the inside. Their corporate owners had compromised the capacity of their networks through ruthless budget cutting as a means of freeing up cash for shareholders. Cargo shipments were stuck at choke points across the country in part because the railroads could not hire enough maintenance workers—a direct consequence of decades of cold-blooded exploitation.

As I spoke with a traveling maintenance crew for a major railroad in the fall of 2022 in Texas, men recounted the pain of being on the road while their children grew up without them. They described disgusting and unsafe lodging. They complained that their bosses refused days off for medical appointments, for the births of children, or for funerals.

"There's no use arguing," one man told me. "We know the deal."

This was a story whose basic elements had changed little since the advent of the locomotive.

Ever since the nineteenth century, when the magnates known as the Robber Barons laid tracks across the American frontier, the business concerns that dominated the industry tended to treat railroads less as a form of transportation than a medium for multiplying their wealth.

Through financial shenanigans, duplicity, and the ruthless pursuit of monopoly power, they prioritized their own fortunes above the interests of their customers, to say nothing of their workers or national prosperity.

In recent decades, the railroads had enhanced their profits by breaching traditional antitrust considerations to engage in a colossal merger spree. They consolidated dozens of once-competing systems into a handful of behemoths with supreme holds over their markets. They applied their dominance toward lifting prices and focusing on their most lucrative routes while abandoning less rewarding swaths of the country.

Not for the first time, the people running the railroads had compromised the resilience of their operations in exchange for sweeter profits.

The railroads were in many ways the quintessential American industry, a showcase of the nation's ingenuity, resourcefulness, and perseverance, along with less-celebrated features like corruption, ruthlessness, and greed.

The locomotive was at the heart of the story of how the United States was transformed from an agrarian country into the most powerful industrial economy on earth, with railroads emerging as the nation's first truly large-scale businesses. Their swift growth and scale necessitated sophisticated new methods of management, from modern accounting systems to the development of corporations comprised of multiple divisions. The spread of rail prompted the adoption of standardized time zones to allow for coordinated scheduling.

Magnates like J. P. Morgan, Cornelius Vanderbilt, and Jay Gould constructed their fortunes in large part through the building and financing of railroads. The monopoly power they amassed became an impetus for federal intervention in the rough-and-tumble American marketplace.

The first rail line constructed in the United States extended from the port of Baltimore to the Ohio Valley, making its inaugural run in 1830. It linked the cities and towns of the eastern seaboard to the bountiful soils to the west, providing a far swifter alternative to the Erie Canal. Soon, men were laying track across the vast expanses of the continent at a fevered pace, opening up the American frontier for all manner of exploitation, from farming and ranching to mining and trading.

The locomotive supplied the cotton plantations of the South with an economical way to ship their crops to the clattering textile mills of New England, boosting profits secured through slave labor. After the Civil War, formerly enslaved Black Americans rode the rails to cities in the North.

Rail transportation carried people of European descent from cities and towns on the East Coast into the hinterland, where they colonized Native American lands, dividing what had been boundless ranges for buffalo herds into delineated parcels of farmland. When some tribes responded by attacking rail construction sites and workers, the US Army used trains to move in troops and supplies so it could subjugate the Native communities.

Trains carried farming equipment west, turning the plains into some of the most productive agricultural lands on earth. Each advance laid the ground for the next, until the rail corridors filled with towns and cities.

From the beginning, the possibilities of train travel intoxicated the investor class. Here was a technological marvel that had the capacity to obliterate the traditional confines of geography.

Asa Whitney, the New Yorker who first promoted the idea of a transcontinental railway, took inspiration from his first train trip in 1844, having just returned from China on an excruciating ocean voyage that consumed five months each way. The contrast between that journey and the pace of the rails left him awestruck.

"Time and space are annihilated by steam, we pass through a city a town yea a country, like an arrow from Jupiter's bow," he wrote in his diary. "As I pass along with lightning speed and cast my eye on the distant objects, they all seem in a whirl."

In the middle of the nineteenth century, the 363-mile journey from Albany, New York, to Buffalo could take up to four days via the Erie Canal. The same trip by rail soon required only five hours.

The construction of the transcontinental railway riveted the nation and the world, much like the excitement triggered a century later by humans setting foot on the moon. For workers, the rigors of the project were extreme and perilous—especially on the 1,800-mile-long western leg overseen by the Central Pacific Railroad, which extended from Sacramento into the Sierra Nevada. They had to blast tunnels through solid hunks of granite and erect trestles spanning canyons.

The project began in 1863. By the following year, with the dangers of the undertaking increasingly stark, the overwhelmingly white workers were quitting in droves. Faced with a critical shortage of labor, the Central Pacific began recruiting Chinese immigrants from surrounding mining camps. Soon, Chinese laborers numbered more than twelve thousand, making up some 90 percent of the workforce on some sections of the line.

The vital role of Chinese workers in constructing what was then the largest infrastructure project in American history established the template for the railroads' treatment of laborers for generations after. The imperatives of the bottom line trumped all other considerations, subordinating respect for human dignity.

Chinese workers were in the United States by dint of an organized recruitment system run by American companies that were eager for low-wage labor. The debts many incurred to arrange their transport across the Pacific made them desperate for income. They were willing to work longer hours and for lower pay than their European counterparts. They proved instrumental in allowing the Central Pacific to cut its wage outlays nearly in half.

The railroads treated Chinese laborers like replaceable parts, frequently not bothering to record their names in company logs, assigning them the riskiest tasks, and relegating them to crude canvas tents in camps segregated from the rest of the operations.

On job sites, Chinese workers contended with an unending litany of hazards, from deadly cave-ins to rapidly spreading diseases like smallpox. They worked in subzero temperatures and in blistering heat while facing the constant threat of racist-fueled mob violence.

The railroads typically provided no medical care, to say nothing of basic lodging, food, or security. When their Chinese employees died, the railroads boxed up their remains and put them on trains headed west, toward ships that would carry them back across the Pacific.

For railroad executives, the vulnerability of Chinese workers—their status as outsiders in a hostile land—was precisely the appeal of employing them. They were insurance against labor agitation.

"A large part of our force are Chinese and they prove nearly equal to white men in the amount of labor they perform, and are far more reliable," a Central Pacific attorney wrote to a member of Congress. "No danger of strikes among them."

The transcontinental line was completed with the ceremonial hammering of a golden spike into the final section of track at Promontory, Utah, on May 10, 1869. As word of the achievement spread by telegraph, people took to the streets in cities across the country. Lower Manhattan shuddered with celebratory cannon fire. In Chicago, tens of thousands of people formed a parade that extended for seven miles.

But the exultation ignored one key feature of the achievement. More than a testament of American engineering prowess, the railroads were a triumph for the powers of the rapacious financiers who ruled the era. They laid down a template that has endured to present times, with their interest in profiteering eternally eclipsing their incentive to provide reliable transportation.

The Robber Barons courted capital from as far away as Europe to finance new undertakings, peddling stories that catered to popular fascination with the locomotive as an agent of progress so potent that it obviated the need for arithmetic.

"The explosion of railroad building across the West could seldom be justified by existing passenger or freight traffic, for the lines traversed a wasteland," wrote the journalist Michael Hiltzik in *Iron Empires,* his gripping chronicle of the escapades of the Robber Barons. "The promoters exploited a popular craze for railroad shares that far outstripped rationality or practicality, but was based instead on the new technology's supposed potential to transform barren territories into bustling Edens."

Union Pacific presented a pungent example of how the unscrupulous pursuit of short-term profits warped the operations of the railroads as the construction boom unfolded in the middle of the nineteenth century. The company constructed the other half of the transcontinental railroad, running west from Iowa.

The federal government was then encouraging companies to build out rail systems as a spur to commerce. It was handing out grants of land running alongside newly installed lengths of track along with long-term loans. Seeking as much free real estate as possible, Union Pacific laid down meandering routes that lengthened rail journeys, trading efficient transportation links for extra land grants. It built sections of track as quickly as possible, leading to shoddy work that yielded frequent breakdowns.

The investors behind the railroad created a second company with the exotic name of Crédit Mobilier to handle all the construction work. Here was a den of thievery on a monumental scale. Crédit Mobilier overbilled extravagantly on its construction projects, pillaging the Union Pacific, with the bill collected via subsidies from the ultimate sucker—the taxpayer. These transactions were enabled by the comprehensive sprinkling of shares through powerful government offices, winning the tacit cooperation of authorities on the take. Vice President Schuyler Colfax was in on the racket. So was the Speaker of the House, the heads of myriad influential committees on Capitol Hill, and a future president, James Garfield, then a member of Congress from Ohio.

Muckraking journalists exposed the scandal in 1872, turning Crédit Mobilier into shorthand for the perfidy that accompanied the railroad boom. The ruse had succeeded by astonishing proportions, funneling some $20 million—the equivalent of nearly half a billion in today's dollars—into the coffers of connected participants. The legacy of the con was to leave the railroad saddled with impossible debts.

On the heels of the Credit Mobilier scandal came a financial crisis known as the Panic of 1873. It began with a series of cascading bank failures in Europe and quickly spread to the United States. At the center of the unraveling was the loss of confidence in bonds issued by railroads.

In the words of Cornelius Vanderbilt, the robber baron whose own railroad holdings were pulled down by a sharp retreat in share prices, "Building railroads from nowhere to nowhere at public expense is not a legitimate undertaking."

As investors yanked what was left of their money out of the railroad industry, construction ground to a halt, sending laborers scurrying for work. The economic slowdown destroyed demand for hauling goods, triggering cutthroat price wars that consigned more than half of American railroads to bankruptcy within a mere three years.

All of that laid the ground for the next advance in the history of American rail. The survivors of the crisis—especially J. P. Morgan—picked through the wreckage to snap up formidable holdings at bargain prices. They merged rail systems and reduced prices, consolidating them to eliminate competition.

Morganization entered the American business lexicon, a term that described what happened after savvy financiers seized control of failing concerns. They fired doddering old managers and replaced them with pliant people. They ditched loss-making routes and ruthlessly pursued monopolies.

From an investor perspective, overseers like Morgan were professionalizing the railroads, eliminating waste and corruption, and imposing order. Once they eliminated competition, establishing chokeholds on major industrial and agricultural centers, the railroads were free to lift prices and boost profits.

Yet from the vantage point of farmers and merchants who depended upon trains to move their crops and wares, here was an organized assault on their livelihoods. By the late nineteenth century, populist movements were drawing mass support by mobilizing to take aim at the railroads, coalescing in statehouses to demand regulation of rates.

And for rail workers, the Morganites represented a direct attack on their paychecks. The streamlining of services and destruction of competition came accompanied by cuts in wages. The result was an extraordinary eruption of social unrest.

"The industrial worker became an unskilled cipher, an interchangeable unit of labor, competing with hundreds of thousands of equally usable men throughout the nation," wrote the historian Robert V. Bruce.

For working people, the emergence of corporate management diluted the power of interpersonal relationships in their dealings with their employers. Their working conditions and their pay were no longer determined by those who hired and supervised them at job sites, but by unseen executives who ruled from the safe remove of their suites in New York, Baltimore, and Philadelphia, and who catered above all to the demands of investors.

Dividends for shareholders and the well-being of workers were locked in a zero-sum competition, limiting spending that might have otherwise tempered the dangers of the rails. Worst off were the brakemen, who clambered across the roofs of railcars and down to the next, manually applying brakes to each carriage. At any moment, they might be decapitated by a low-slung bridge or crushed by rolling wheels after tripping and falling. Winter intensified the risks, bringing the prospect of sliding off an ice-encrusted sheet of metal and landing beneath the moving train.

"A brakeman with both hands and all his fingers was either remarkably skillful, incredibly lucky or new on the job," wrote Bruce. "Most railroads waited years before installing well-known safety devices, because, the men bitterly assumed, condolences came cheaper."

Yet, in what remained a consistent stance through the present day, the railroads withheld support for those who suffered illness—even when workers were injured on the job.

"The regular compensation of employees covers all risk or liability to accident," declared the official regulations governing the Pennsylvania Railroad, the richest and most powerful of the lot. "If an employee is disabled by sickness or any other cause, the right to claim compensation is not recognized."

Rail workers resented the punishing demands of their schedules, and especially the layover—the time train crews had to spend at the end of one journey waiting for the next, while paying for their own lodging and meals. The alternative was to ride home by train at their own cost. Workers were expected to be available whenever the railroad needed them, though they were paid only for their time on the job.

These were the circumstances governing the lives of railroad workers before their employers began cutting their wages in earnest in the spring of 1877.

The Pennsylvania slashed pay by 10 percent, with the cuts taking effect on June 1. Three days later, the railroad's president, Thomas A. Scott, met with aggrieved workers who implored him to reconsider. They got nowhere. Scott, in fact, lamented his own troubles. The panic had forced the railroad to reduce its dividend. It was only fair for rank-and-file workers to take on some of the burden, he said. Scott left out the fact that the Pennsylvania had only the previous year found the money to distribute 8 percent of its earnings in dividends while still managing to accumulate a cash hoard of $1.5 million.

Scott's appeal to the spirit of shared sacrifice added fuel to what exploded into one of the great conflagrations in American labor history. As other railroads in turn cut wages, Pittsburgh became the epicenter of a national wave of strikes and demonstrations.

As enraged crowds of workers poured into the Pittsburgh rail yards, they halted trains and monkey-wrenched locomotives, rallying around a simple demand twinned with a threat. "We will have bread or blood," they chanted.

Scott's retort neatly summed up the enduring treatment of workers by American railroads for generations after. The strikers should be administered "a rifle diet for a few days and see how they like that kind of bread."

The railroad boss made good on that threat. After local men deployed to quell the strikers took the side of the workers, Scott persuaded state authorities to bring in militia forces from Philadelphia, on the other side of the state.

As six hundred troops arrived in railcars on the afternoon of July 21, they absorbed the spectacle of two thousand train cars and locomotives sitting motionless. Water from melting ice seeped from refrigerated cars full of meat rotting in the midsummer heat. Heaps of fruit and vegetables sat decomposing.

Waylaid in Pittsburgh "were all the necessities and most of the luxuries of nineteenth century civilization," wrote Bruce. "Clothing, furniture, books, whisky, silverware, oil, coal, flour, machinery, carpets, ornaments—everything from cribs to coffins."

The Philadelphia militia had come armed with rifles and a pair of Gatling guns, the forerunner to the machine gun. As the men marched in neat columns to confront the strikers holding up what we now call the supply chain, a crowd of more than five thousand people looked down from a hillside. Some taunted the troops, and a few threw stones and hunks of soil. The militia responded with a scattershot barrage of fire.

The shooting killed more than two dozen people and severely injured many more. As word of the massacre spread, fresh crowds descended on the rail yards. People set fire to railcars loaded with oil and sent them hurtling toward the depot that had become a refuge for the Philadelphia militiamen. The inferno was so enormous that communities a dozen miles removed from Pittsburgh saw an orange glow on the horizon.

The strikes went on and the flow of cargo remained disrupted, until President Rutherford B. Hayes unleashed federal troops to put down what he described as an "insurrection." He was responding to the pleas of governors, who were themselves besieged with demands for intervention from railroad bosses.

In his diary, Hayes recorded his feeling that the labor turmoil was merely the surface manifestation of a deeper set of ills, a rottenness that had taken hold of the American economy. It was unduly influenced by the interests of the business magnates who controlled the railroads.

"The strikes have been put down by force; but now for the real remedy," he wrote. "Can't something [be] done by education of the strikers, by judicious control of the capitalists, by wise general policy to end or diminish the evil? The railroad strikers, as a rule, are good men, sober, intelligent, and industrious."

On a blazing hot afternoon in October 2022 in a desolate stretch of Texas, a half dozen railroad workers stood in a parking lot outside a grungy motel and described why they and tens of thousands of their union brethren had grown angry enough to walk off their jobs. Their union had just voted to authorize a strike.

Nearly a century and a half had transpired since the great upheaval of 1877. Yet among rail workers, core grievances retained striking similarities.

The men—they were all men—were enraged that they lacked paid sick leave. They were disgusted by their meager reimbursement for food and lodging while on the road, forcing them to cram two and three to a room into dilapidated motels rife with bedbugs and rats.

Above all, they resented the railroad's callous disregard for the burdens of their lives—their need to manage their jobs alongside their responsibilities to their families.

The men were members of a traveling maintenance gang for a major American railroad. They ranged across the midsection of the United States, repairing aging lengths of track and attending to emergencies.

They had driven from five to thirteen hours to get to their job site in Texas, leaving behind homes as far away as Louisiana, Oklahoma, and Mississippi. They had been forced by a last-minute change in their schedule to arrive a day earlier than planned. They had just learned that they would have to remain there for an extra day, cutting their time at home before their next assignment.

One man had recently had surgery and was due for a scan to monitor his condition. He had scheduled an appointment for what had initially been a week off. When the railroad abruptly changed the schedule, he had been forced to delay it.

The men were accustomed to the gnawing regrets of life on the road, regularly missing wedding anniversaries, children's birthday parties, graduations, and the deaths of parents. They were constantly disappointing the most important people in their lives.

"To have a successful family, you've got to have a wife who's very understanding," one man told me. "If not, you ain't gonna be married long."

They were also conditioned to laboring even when they were ill, and frequently foregoing doctor's visits, because their contract with the railroad—a contract that had run out three years earlier—provided no paid sick leave.

They were so fearful of their employer that they agreed to speak to me only on the promise that I not identify them by name or reveal their hometowns. People had been fired for simply naming their employers in Facebook posts, they told me. Their bosses were mercilessly intolerant of criticism.

"These people," he said, referring to the railroad bosses, "are the same as two hundred years ago. These people do what they want."

Much had obviously changed since the heyday of the Robber Barons, from the advent of modern safety equipment to pay that frequently reached six figures, secured through collective bargaining. And yet much had indeed remained the same.

The populist movements of the nineteenth century had drawn strength from the ire of farmers and merchants aggrieved by the price gouging of monopolistic railroads. That coalesced into a substantial victory in 1887, as Congress created the Interstate Commerce Commission, a federal body tasked with regulating the railroads. Three years later, Congress unleashed another assault on monopoly power in the form of the Sherman Antitrust Act.

The regulations that stemmed from these foundations in American antitrust law were riddled with loopholes and inconsistently enforced. The ICC banned price-fixing among railroads, but that merely spurred a wave of consolidation to get around that stricture. Two competitors were barred from coordinating their prices. But the same two companies could do as they pleased within the confines of the single company yielded by a merger.

Still, the decades that followed were characterized by greater stability under the gaze of empowered government watchdogs. Railroads had to publicize their prices and offer the same terms to all customers without discrimination, much as the ocean carriers did before the undoing of the shipping act.

Then, in the middle of the twentieth century, technological change delivered a fresh jolt. First, Henry Ford turned the family automobile into a commonplace item. By the mid-1950s, President Eisenhower embarked on the construction of the interstate highway system. In the business of moving freight, the railroads had a new and nimble competitor—the long-haul trucking industry.

Trucking grew rapidly at the railroads' expense. From the end of World War II until 1975, the share of American cargo carried by rail dropped roughly in half—from nearly 70 percent to 37 percent. At the same time, an aggressive ICC treated the railroads like a national utility system, pressuring them to serve even less profitable routes. This proved devastating to the bottom line. By the late 1970s, more than one-fifth of the American train system was being operated by railroads ensnared in bankruptcy.

This was the backdrop to another pivotal act of deregulation: in October 1980, President Jimmy Carter signed into law the Staggers Rail Act, freeing the railroads from key elements of federal oversight, and entrusting a critical component of the supply chain to corporations left to their own devices.

"By stripping away needless and costly regulation in favor of marketplace forces," Carter declared, "this act will help and assure a strong and healthy future for our nation's railroads and the men and women who work for them."

Carter was describing as one and the same the interests of the railroads and those of their workers, a notion that collided like an onrushing locomotive with the lessons of history.

He also promised that deregulation would deliver a bonanza for American businesses reliant on the rails to move freight. "It will benefit shippers throughout the country by encouraging railroads to improve their equipment and better tailor their service to shipper needs," the president said.

Here was another dubious proposition. The story of the railroads over generations was a tale of profit maximization coming at the direct expense of resilience. A few people warned that Carter's deregulation would reprise that effect.

"We will have fewer lines, fewer railroads providing less service at higher cost," said Representative Henry B. Gonzalez, a Democrat from Texas.

Those words proved prescient.

The railroads used their newfound freedom to bargain privately and set prices absent the ICC's approval. They gave better rates to the largest players. This was a major catalyst for the rise of retail giants like Walmart and Home Depot, who gained another advantage over smaller competitors—the ability to move their goods at lower cost.

The railroads were also liberated from their obligations to serve every stop along the way. With the ICC weakened and eventually disbanded, rail networks began dropping routes as suited their ultimate masters—their shareholders.

Over the first three decades that followed Carter's signing of the Staggers Rail Act, the volume of freight hauled by the railroads increased by roughly a third even as they reduced their miles of track by 40 percent. The consulting crowd hailed an increase in "traffic density"—more stuff being moved over smaller networks. This was an ostensible marker of efficiency that lifted railroad stock prices.

But this efficiency came accompanied by costs that were ultimately inefficient for those who depended on rail transportation.

The wounded remains of the ICC approved every railroad merger so long as it left at least two competitors on every route. After the ICC was put out of its misery in 1995, handing its remaining regulatory authority to the freshly created Surface Transportation Board, the new body continued on in that spirit.

The result was to return the railroads to their historic status as a preserve for monopolists.

By 2014, the number of so-called Class 1 railroads—the government's designation for the largest systems—had dropped to a mere seven, down from thirty-three when the Staggers Rail Act became law. Over that same thirty-four-year span, the four largest railroads increased their share of the market for moving grains and other farm products from 53 percent to 86 percent.

In much of the country, the rail freight market became a duopoly, with only two competitors on most routes. At more than three-fourths of the freight stations in the United States, only one rail line provided service, the very definition of a monopoly.

Though rates dropped during the first fourteen years of the Staggers Rail Act, they rose substantially over the following two decades. By 2016, the cost of rail freight had increased by more than half compared to 2004, according to one analysis.

Not coincidentally, profits for major railroads nearly tripled between 2000 and 2017, even as their costs increased by only 3 percent.

And then came another surge in profitability via the railroads' latest financial innovation. They reduced service and laid off workers to cut costs while giving the resulting cash to shareholders.

And they draped the proceedings with a canny name that lent their activities a whiff of scientific rigor and management discipline: Precision Scheduled Railroading.

CHAPTER 14

"THE ALMIGHTY OPERATING RATIO"

Modern-Day Pillaging of the Rails

Precision Scheduled Railroading applied the Just in Time logic to rail freight.

In the years leading up to the pandemic, six of the seven largest railroads in the United States implemented a version of it. They touted Precision as a way to better serve customers by tailoring train schedules to the needs of shippers. They celebrated a systematic approach to boosting efficiency by ruthlessly eliminating waste.

In reality, Precision was another bit of corporate argot propagated by the consulting class to justify shifting compensation from ordinary workers to executives and investors.

Under the Precision banner, the railroads reduced spending on locomotives and maintenance equipment. They slashed overtime. And they relentlessly eliminated jobs while transferring the savings to investors. Between 2016 and 2022, the largest railroads cut their workforces by 29 percent, or some forty-five thousand people.

Not for the first time, gains for investors came at the direct expense of railroad service.

Despite the promises of more predictable schedules, the railroads frequently delayed trains until they amassed enough cargo to optimize the profits on any individual run, stretching out strings of cars that exceeded three miles long.

Meanwhile, the remaining employees were left to shoulder greater burdens at reduced pay. They worked longer hours, endured more time away from their families, and suffered greater risk of accidents. Customers contended with breakdowns, delays, and uncertainty.

"I can only conjecture what is driving all this, but all parties know, and are being driven by something, something that needs to be reined in," wrote a railway worker in Iowa, in an open letter to the federal body that regulated the railroads, the Surface Transportation Board. "It is Wall Street greed and investor demands. Sixteen-thousand four-hundred and fifty-foot trains weighing more than 42 million pounds are gratifying someone with power. Someone who wants it all and more."

Union Pacific had been a conspicuously enthusiastic proponent of Precision Scheduled Railroading. Not coincidentally, its network proved especially incapable of handling the strain from the surge of containers reaching American ports.

This helped explain why Hagan Walker wisely put aside any thought of shipping his box from Southern California by train. He knew little about the rail system or the intricacies of cross-country freight. But in every conversation with someone in the logistics business, railroad freight was quickly dismissed as an invitation for cascading troubles.

Back in the spring of 2019, Union Pacific had closed one of five so-called ramps—the entry ports for cargo—at a major railyard in Chicago. The railroad's executive vice president for marketing, Kenny G. Rocker, presented this as a critical simplification that would "allow us to streamline operations and offer faster loading and unloading of containers," which would deliver "benefits to our customers."

The same year, Union Pacific boasted to investors that it had slashed its locomotive fleet by one-fifth. Its freight volumes dipped by 6 percent that year, but the company still managed to pay out $2.6 billion in dividends—more than a third of its earnings.

Two years later, in July 2021, as imported goods swamped American docks, Union Pacific halted shipments of containers traveling east from West Coast ports. The railroad described this measure as unavoidable given "significant congestion at our inland intermodal terminals, most notably in Chicago."

That move revealed the ruse. The action plan that had accompanied the railroad's rhetorical reach for simplification, efficiency, and precision had produced none of those things. Rather, it had generated new vulnerabilities. Much as consultancies like McKinsey had warped the sensible thinking behind Toyota's concept of Just in Time, the railroads had deployed Precision as the justification to ransack their own operations for the enrichment of shareholders.

In late October 2021, just as Glo's container was off-loaded onto the docks at the port of Long Beach, Union Pacific's system was buckling. Its rail yards in Southern California were so inundated with empty boxes that it would no longer accept them for shipment. It was turning away containers coming in from Chicago and points to the east, leaving them stranded in the middle of the country.

This was part of the reason why almond growers and other agricultural exporters in California's Central Valley could not secure containers needed to ship their goods. Union Pacific and other major rail networks were not equipped to handle the volume of freight confronting them, so they abandoned less profitable pursuits like hauling empty containers.

And this was partly why store shelves were short of everything from sporting goods to cabinetry, and why prices were soaring for consumer goods, prompting central banks to lift interest rates in an aggressive bid to snuff out inflation. The railroads were not geared to manage a surge of demand.

But if Precision was exposed as a failure in terms of efficiently moving freight, it succeeded dramatically in turning railroad stocks into more alluring investments. Specifically, Precision lowered the railroads' so-called operating ratios, meaning their costs of running their systems as a percentage of their total revenue. Wall Street was especially fixated on that metric, rewarding progress with higher share prices.

In the same way that manufacturers and retailers had ditched inventory to reduce assets on their books, thus bolstering their return on assets, railroads used *precision* to goose their numbers.

Another metric that had become a Union Pacific obsession was dwell time—the period that a particular container sat in a rail yard. Managers were so intent on driving down the average dwell time at their yards that they sometimes sent cars in the wrong direction, bound for terminals where they would become someone else's problem, rather than hold up trains.

In July 2022, a Union Pacific engineer in Idaho, Michael Paul Lindsey II, discovered that a train he was driving had been loaded with scores of cars that had been misrouted. The train had originated at a yard in Nebraska and was headed west to Oregon. It was carrying diesel fuel bound for Idaho that had to be unloaded in Oregon and then put on another train headed back in the opposite direction. It was pulling racks full of automobiles destined for California that should have been attached to another train headed there.

"How many days or weeks of additional transit time will this cause for the businesses and customers that need these vehicles to arrive on time?" Lindsey wrote in a letter he sent to the Surface Transportation Board. "This bothers me immensely."

The culprit was Precision, Lindsey concluded. Surely the terminal operators in Nebraska knew that they were dispatching cargo toward the wrong destinations. But they were governed by the compulsion to keep freight moving. Better to attach cars to a train leaving right now—any train, going anywhere—than allow them to sit in the yard, increasing the all-important dwell time, which would hurt the stock price.

Lindsey had grown up dreaming of working on the railroad, captivated by the romance of trains. Yet somewhere between being away from his family on more Christmases than he could remember and the collapse of his marriage, he had come to contemplate other options.

He had initially seen upsides to Precision, viewing it as an analytical exercise in rooting out waste, much in the spirit of Toyota's landmark mission. But as he witnessed its impacts on the trains he was driving, he concluded that Precision was mostly about satisfying Wall Street's yearning for whatever performance indicator was making stock prices go up.

"It's transitioned to 'just screw the railroads,'" Lindsey said. "'Screw our future growth, we just want to squeeze as much profit out of the railroads short-term as we can.'"

Union Pacific formally adopted Precision in October 2018. Three years later, its share price had nearly doubled, boosting stock-based compensation for its executives. The railroad's chief executive officer, Lance M. Fritz, would take home more than $14 million in 2021. Only $1.2 million of that was in salary, while $10.5 million came via grants of stock and stock options.

Rocker, the executive who had celebrated the shutdown of capacity in Chicago, collected more than $2.7 million in compensation that same year.

Shippers made out less spectacularly.

As the pandemic brought a surge of demand that overwhelmed the American freight system, chemical companies with factories along the Gulf of Mexico complained that they could not get their products onto trains. That left plants on the East Coast short of critical ingredients, impeding their ability to make a vast range of goods, from pharmaceuticals to paint and industrial solvents.

"Rail service for chemical shippers continues to deteriorate," declared Chris Jahn, the president of the American Chemistry Council, in testimony to the Surface Transportation Board in April 2022. He disclosed that Union Pacific had gone so far as to ask customers to reduce their cargo because it could not handle the flow.

"Years of railroad decisions to cut staff, eliminate switch yards, and slash customer service resources have gutted network resilience, making service crises like this one entirely predictable, if not inevitable," Jahn said.

The fertilizer industry, which shipped more than half its goods by rail, was recounting similar nightmares—service cuts; cargo rerouted to distant points to bypass congestion; delays given shortages of train crew. As a result, farmers were paying higher prices for fertilizer, making food costlier for American households.

In Kansas City, Kansas, where General Motors churned out Cadillac SUVs and Chevy Malibu sedans, unreliable rail service was preventing the automaker from shipping many new vehicles to dealerships, contributing to shortages and soaring prices.

The GM plant had shut down for several months in the face of chip shortages but was struggling to resume normal operations when I visited its shipping terminal on a brutally cold morning in November 2021.

The automaker deposited freshly completed cars across a sprawling parking lot. Its terminal was run by an outside company called Jack Cooper, which handled the logistics of getting these cars to points across the nation, sending some by truck and more by rail.

The man in charge of the shipping terminal, David Heide, was accustomed to hearing from his counterparts at GM, often many times a day, demanding to know when the latest batch of vehicles would go out.

"They're always banging on us about moving more product," Heide told me. "And we're saying, 'We're probably not going to load as many as you think we are.'"

Born and raised in the middle of Kansas, Heide navigated life like the catcher he had been on his college baseball team. He walked the terminal with the easygoing confidence of someone who knew his way around, greeting everyone and accepting perpetual ribbing while quietly delivering orders. But his joviality did not mask his intensity, or his simmering frustration over having to produce results in a system dominated by factors that were beyond his control.

His most consistent source of frustration was never knowing whether the railroads would deliver as many railcars as they promised.

The previous Friday, the railroads had failed to set up any cars in his yard, leaving several hundred GM vehicles piling up in the lot. By Monday, GM was pressuring him to move them out as quickly as possible. But Heide was reluctant to summon workers for the job until he was sure the railcars were in place.

"They want you to call in a whole crew to load the train," Heide said. That would leave him paying full-time wages for what was likely to turn out to be only three days of work when, inevitably, the railroad failed to deliver enough cars.

The alternative—summoning a full crew and then sending people home—was demoralizing.

"You don't want to call people back and yo-yo them," Heide said.

The Malibu sedans were loaded into railcars that were stacked to three levels, but the larger Cadillac SUVs required two-level cars. Heide could never get a straight story on which variety of railcars were on the way to his yard and when.

"There isn't any consistency to inbound rail," he said.

In Washington, the Surface Transportation Board was besieged by such complaints, prompting its chairman, Martin J. Oberman, to strike an unusually combative stance.

In a blistering speech delivered that fall, Oberman noted that major American railroads had collectively bestowed "an astounding $183 billion" on shareholders over the previous eleven years via buybacks and dividends. This money had come at the direct expense of investments that could have improved rail service, effectively bilking shippers, rail workers, and the American economy.

"Through ever increasing pressure from Wall Street, the railroads' emphasis has not been on growth," Oberman said. "Rather the emphasis has been on cutting in pursuit of the almighty operating ratio."

The following May, Oberman accused the railroads of sacrificing their capacity to move freight on the altar of *lean*.

"The rail industry clearly is struggling to provide adequate and reliable rail service," Oberman said during testimony on Capitol Hill. "They have cut labor below the bone. In order to make up for the shortage of labor, they are overworking and abusing the workforces they have."

By then, the railroads were promising to rectify their labor shortages, touting aggressive recruitment efforts that would plug the gap and restore order.

"Service is not yet where we want it to be," the president of Norfolk Southern, Alan Shaw, told investors during a call detailing the railroad's latest earnings in July 2022.

The railroad's chief operating officer, Cindy Sanborn, offered assurances that more workers were on the way to trouble spots. "We're taking advantage of every option to get folks where we need them," she said.

There was one problem with such talk.

The word was out that working for the railroad was a miserable existence. Precision had downgraded the job to such an extent that even longtime workers were quitting.

For Anthony Gunter, the final indignity came in the summer of 2022.

His employer, Norfolk Southern, was congratulating itself for record revenues at the same time that it was arguing that it could not afford to provide paid sick leave to workers like him.

At his home in eastern Tennessee, Gunter sat down at his computer and typed an email to his supervisor.

"Due to the corporate greed and the extreme disrespect that this company has shown," he wrote, "I'll be hanging up my hat just shy of fourteen years. Norfolk Southern has no recognition for good employees that show up to work on time and do their job safely, efficiently and take the extra initiative to go above and beyond production expectations!"

Gunter did not walk away lightly. His father had worked for the same railroad on a traveling maintenance gang for more than forty years. Gunter assumed that he, too, would endure that life until his own retirement, even as he recalled the pain of watching his father leave when he was a child, sometimes sneaking into his dad's duffel bag to try to follow him out on the road.

The job was rough, but the money was unbeatable in his stretch of rural America.

After graduating from high school in 2008, Gunter joined a Norfolk Southern maintenance gang that roamed from Georgia to Illinois. He worked four ten-hour shifts in a row, driving as far as twelve hours to get to and from his home in Tennessee to his job sites.

He swung giant hammers, pounding stakes into railroad ties. He moved metal plates weighing more than twenty pounds, bending over and dragging them by hand, nursing aching muscles every night.

He slept in camp cars set alongside the tracks—metal boxes lined with bunk beds packed so tightly he could reach over and graze the man sleeping next to him. The showers were frequently out of order. The toilet was portable. Meals provided by the railroad were prison grade—shriveled hot dogs, baked beans, gristly hamburger patties.

But that first year, he earned $18.87 per hour plus overtime, bringing home $55,000. By 2015, he was making $86,000 a year.

Then the railroad implemented Precision.

Suddenly, Gunter and his crew were forbidden from working

overtime, even as they were expected to complete projects on an accelerated schedule—more pressure for lower pay.

The railroad stopped paying workers for the first half hour they spent in company vehicles riding from camp to their work sites and back. This lopped off an hour of pay per day, enraging Gunter and the others in the gang.

By 2017, his take-home pay had slipped to $70,000. That same year, he got married. Two years later, his daughter was born, and Gunter gained an up-close understanding of what his own parents had weathered—the physical distance, the frustrations of the spouse left to take care of children alone, and the difficulties of returning home.

"It was definitely a strain on the marriage," he said.

In 2021, his son was born with a heart defect, requiring surgery eight months later. Gunter stayed home for the baby's stint in the hospital, resigned to foregoing pay. Even then, his supervisor pressured him to return.

"He said, 'You're putting me in a tough spot,'" Gunter recalled. "'We're short-handed this week. You have to be here. We can't have you leaving.'"

He struggled to make sense of this even a year later, his voice trembling with anger.

"They were making me feel like crap because I wasn't there," Gunter said. "It's not a good way of life. It's almost like a brainwashing cult that they make you think that you should never be anything better than what you are right there. That you can't improve your life. That you're not worth more."

That summer, his union and eleven others collectively representing more than one hundred thousand rail workers nationwide began mobilizing toward a potential strike. They sought to break an impasse in talks with the railroads on a new contract.

Two and a half years had elapsed since their previous contract had reached its end, meaning they were working without a raise, even after enduring the pandemic. Resentment within the ranks was mounting. Yet their right to cease work was constrained by federal laws that went back to the upheaval of 1877, proscribing a series of negotiating stages that had to be exhausted before they could legally strike.

For President Joe Biden, here was a dangerous and volatile situation.

Midterm elections were approaching, animated by discontent over inflation. A strike that shut down the American rail system would dramatically exacerbate the supply chain crisis, making this an outcome to be strenuously avoided. Yet Biden claimed credentials as a stridently pro-union president, an image that made him reluctant to impose a settlement.

Seeking middle ground, Biden convened a presidential emergency board to make recommendations that could provide the basis for a deal.

In testimony to the presidential board, the unions noted that their workers had been laboring without a hitch during the pandemic. The chief executives of three of the largest railroads— among them Norfolk Southern and Union Pacific—had taken home $183 million between 2018 and 2021, with their compensation more than doubling. During those same years, worker pay had increased by less than 14 percent.

In a phrase that quickly became a rallying cry among workers, the railroads dismissed the relevance of the pay disparity. "The carriers maintain that capital investment and risk are the reasons for their profits, not any contribution by labor," read a section of the board's report.

"We thought it was bullshit, but with a few more words," one of the guys in Texas told me.

Another man rolled his eyes. "If that's true, then why don't they let us strike?" he said. "If we don't make you no money, then why are we here every day?"

The presidential emergency board detailed recommendations for the terms of a settlement: wage increases of 24 percent over a five-year period, lifting average pay to $110,000 a year, plus annual cash bonuses of $1,000. The railroads pointed at those numbers as indications of the most generous pay increases in half a century. The unions argued that the terms amounted to a pay cut after accounting for inflation.

Of greatest consequence, the presidential board accepted the railroads' assertion that they could not afford to add sick pay without jeopardizing investment into their networks. As a compromise, the board suggested the addition of a single personal day, an addition so meager that workers felt mocked.

Even during the worst of the pandemic, workers regularly showed up sick. It was either that or risk falling behind on their bills. "You had guys that just didn't want to share that they had COVID because they couldn't afford to take a day off," a member of a maintenance gang in Alabama told me. "It added to the spread on the road."

In the midst of contract negotiations, Gunter heard the talk from his employer that its coffers were teeming, even as it brushed off demands for paid sick leave.

"We achieved record revenue for the quarter of $3.3 billion," the railroad's chief marketing officer, Ed Elkins, told stock analysts.

The railroad's chief financial officer, Mark George, warned that "the current labor negotiations" could result in "incremental headwinds" to the company's bottom line.

Gunter was incensed. He sent in his resignation letter. Within three minutes, he received a phone call from a corporate vice president wanting to know if he was really leaving.

"He said, 'We're losing a lot of good men,'" Gunter recalled. "'Are you sure?'"

He was sure. He already had another job lined up, working construction at a nuclear plant close to home. He'd make more money and be able to spend far more time with his family. Still, walking away from Norfolk Southern was emotional.

"It made me mad and sad at the same time," he said.

———

Henry Ford had seen this one coming, too.

He had worried that pressures from shareholders would interfere with his ability to pay robust wages. And that would menace his ability to attract enough workers to assemble record volumes of cars.

Ford was no labor hero. In the end, he sought to destroy organized labor, firing workers who tried to unionize his plants while unleashing brutal violence to suppress their mobilization. Still, in his heyday, he grasped a truth that has been brought home emphatically in recent years. He understood that he had to pay his employees what we today call a living wage or otherwise suffer a loss of morale that would threaten his productivity. Fair recompense was the means of ensuring that people showed up to work undistracted by stresses over bills.

A demanding boss, he was cognizant that sustained and focused attention was something that could not be purchased at a discount.

"No manufacturer in his right mind would contend that buying only the cheapest materials is the way to make certain of manufacturing the best article," Ford wrote in his autobiography. "Then why do we hear so much talk about the 'liquidation of labor' and the benefits that will flow to the country from cutting wages—which means only the cutting of buying power and the curtailing of the home market? What good is industry if it be so unskillfully managed so not to return a living to everyone concerned? No question is more important than that of wages—most of the people of the country live on wages. The scale of their living—the rate of their wages—determines the prosperity of the country."

This was an insight gained from a fundamental problem that vexed mass production: a shortage of willing hands.

Workers on Ford's assembly lines detested their confinement, robotically performing the same tasks over and over. They chafed at being monitored and controlled, with ubiquitous clocks implicitly reminding them of the pressure to maintain a frenetic pace. People quit in droves.

Faced with this threat, Ford responded with a policy that astonished the business world. He doubled the pay of his workers to $5 a day.

That policy, outlined in January 1914, triggered a seismic debate in American life. Some derided Ford as a bleeding heart who was running his business like a charity. The president of a glass company in Pittsburgh accused Ford of courting "the ruin of all business in this country."

But Ford framed the policy in pragmatic terms. He was using the company's money to purchase what it could not do without—skilled and fully engaged labor.

"A low wage business is always insecure," Ford declared.

Ford proffered the raises unilaterally, and not in the context of collective bargaining. As organized labor gained power over subsequent decades, Ford opposed the movement as an enemy of innovation and productivity. The generous pay came on his terms, accompanied by the establishment of an intrusive and creepy "sociological department" that visited workers at their homes to ensure they were not drinking, gambling, or engaging in other unsavory activities—a condition of their employment.

And it bears noting that Ford had the cash to pay high wages because his company held a monopolistic grip over the affordable passenger car market. Generous pay did not come at the expense of returns to shareholders, but rather enabled sustained profitability.

The wage hikes were also engineered in part to bring positive publicity—a public relations gambit that worked with stunning effect. "Social Justice Animates Ford, He Is Not for Multi-Millionaires," blared one sympathetic newspaper headline.

Whatever the motive, the dollars that Ford paid out were real. Laborers massed at Ford plants, swiftly ending worries about recruitment.

In Detroit, a city clerk handed out marriage licenses to fifty Ford workers in the first two weeks after the $5 day was announced—putative evidence that the extra wages had significantly altered the economics of supporting a family. One Ford worker told a local reporter that his son no longer had to sell papers while his daughter was able to quit her domestic job, freeing up more time for gathering.

"Again, we are a family," he said.

Working on the railroad had always been a tough job, and especially for people who had to venture hundreds of miles away from their families. But Precision had clearly made life more difficult.

Because train schedules changed more frequently, so did the schedules of the people who kept them running. And because the railroads had reduced their workforces, they needed the people who remained more than ever. That prompted management to impose draconian attendance policies.

These were among the grievances that compelled four of the dozen major unions to vote down a settlement brokered by the Biden administration.

The White House had announced the deal with much fanfare in September 2022, after late-night talks that had seemed to avert a strike. But the terms struck were largely the same as the recommendations made by the presidential advisory board. The only addition was three days of unpaid leave that workers could schedule for medical visits without being subject to discipline under the new attendance policies.

"There is the potential for a strike," the lead negotiator for the maintenance workers' union, Peter Kennedy, told me just after the rank and file voted down the deal. "It's an insane and cruel system, and these guys are fed up with it."

But the White House would not let a strike happen. Forced to choose between angering the unions and permitting a new supply chain crisis, Biden opted to keep freight rolling. He called on Congress to craft legislation that barred a strike by imposing the terms of the settlement brokered by his administration.

"Look, I know this bill doesn't have paid sick leave that these rail workers and frankly every worker in America deserves," Biden told reporters. "But that fight isn't over."

But the fight was largely theater.

Some Republicans, including Senator Marco Rubio of Florida, expressed support for paid sick leave while vowing to withhold their votes for any settlement that failed to include it. But Biden and House Speaker Nancy Pelosi allowed such rhetorical virtue signaling to go untested. She presented two bills in the House—one that simply approved the terms of the settlement, and another that added seven days of annual paid sick leave as an amendment. By keeping those two pieces of legislation separate, she gave members of Congress an easy way to extinguish the possibility of a strike, while still supplying a performative vote of concern for the well-being of rail workers.

The House quickly approved both measures. The Senate backed the primary bill, thus dispatching the potential for a strike, and sending it to Biden for his promised signature. In a surprise to no one, the Senate failed to muster the sixty votes needed to override a filibuster threat on the issue of sick leave.

In short: no strike, and no paid sick leave.

More than a century after the era of the Robber Barons, the modern-day railroad magnates had won again. And the way in which they had secured victory had inflicted an enduring injury on the rail unions, weakening their leverage in future rounds of collective bargaining.

The powers that be had demonstrated their willingness to intervene and impose whatever terms were required to keep trains moving. They had delivered the message to the railroads that they could hold the line, refusing to satisfy the demands of their workers, and still depend on their labor.

More broadly, the nation took in the latest illustration that any contest between the health of critical workers and the maintenance of the supply chain was no contest at all. The reliable delivery of product took precedence over the welfare of the people who kept the gears turning.

And then came the punch line.

Only four days after Congress voted to force rail workers to continue laboring without paid sick leave, Norfolk Southern convened its annual Investor Day at its corporate headquarters in Atlanta.

Its chief financial officer, Mark George, wandered confidently onstage in a gray suit adorned with a pocket square as he beamed a slide on the wall titled "Returning Capital to Shareholders." A bar chart showed steadily increasing dividends along with the target for future distributions—35 to 40 percent of the company's earnings.

"We have doubled our quarterly dividend in the past five years," George told investors. "We will be growing our dividend more."

Major railroads had persuaded the president and members of Congress that they could not afford to pay sick leave to their workers without jeopardizing their operations. And here was one railroad sharing word with investors that it could afford to hand *them* more than a third of its earnings.

Two months later, just before 9 p.m. on February 3, 2023, a Norfolk Southern freight train pulling 149 cars derailed as it passed through the town of East Palestine, Ohio. Eleven of the cars were full of hazardous materials, including vinyl chloride, a chemical that can cause cancer. As some of the cargo ignited, a thick plume of smoke hovered over the town of 4,700 people. Millions of gallons of toxic liquid poured into surrounding streams and soils.

Federal authorities later identified the immediate cause of the derailment as an overheated wheel bearing. But the broader explanation for the disaster was in part deregulation. As *The Lever* reported, the railroad industry had successfully lobbied the Obama administration to exempt trains like the one that derailed from new rules governing so-called high-hazard flammable trains. The details would have required the addition of upgraded braking technology that might have prevented the derailment. The Trump administration subsequently lifted the rules altogether.

And the clearest culprit was Precision. To gratify shareholders, railroads like Norfolk Southern had entrusted the operation of longer trains than ever in the hands of overworked, underpaid skeleton crews.

"There's tons of people out there that are just exhausted," said Lindsey, the locomotive engineer. "The way the railroad is now, their opinion seems to be that our lives belong to them."

He posted a video to TikTok describing how major railroads had changed a safety system that had previously sent alerts about overheating bearings directly to train drivers. Instead, such information was diverted to central dispatchers, often thousands of miles away, delaying warnings. It was all about keeping trains rolling and minimizing dwell time—a triumph for Precision.

"It's time that Americans ask why giant unaccountable monopolies are allowed to own, neglect, and dominate our vital rail infrastructure while buying back billions of dollars of their own shares," Lindsey declared on the video. "Railroad workers have warned of the cuts to maintenance, longer trains, and utter contempt these companies hold for the general public. East Palestine will happen again."

Three weeks after he posted the video, Lindsey was fired by Union Pacific on grounds of violating their social media policy and divulging sensitive information.

CHAPTER 15

"SWEATSHOPS ON WHEELS"

The Long, Torturous Road

Hager Walker had been correct to assume that he should forget about rail as the means of moving his container from Southern California to Mississippi. He would opt for a truck.

Yet this option entailed its own perils. So-called long-haul trucking companies, which moved cargo over vast distances, were deluged by demand for their services. They could not hire enough drivers to manage the load.

This, too, was a story whose origins went back decades, to another decision by the Carter administration to deregulate. Along the way, truck driving had been downgraded from a tough but middle-class job into a pursuit best avoided.

Walker read the email and felt a deepening sense of dread.

It was November 3, 2021, and the holiday season was only weeks away. His container full of *Sesame Street*–themed Glo dolls sat stranded in a storage yard in Southern California, still waiting for a ride across the country.

The email was from Freightos, the online platform he had used to book the passage from China. It informed him that the delivery date for his container had been updated from October 30—a date already passed—to December 10.

"I said to myself, 'We're going to miss Black Friday and Christmas,'" Walker said.

He wrote to Sunny Liu, the shipping agent in China, beseeching her to help him complete the final leg of the journey. She explained that her role in the shipment was over.

Responsibility for the last segment—the crossing of the continent—fell to a company called Israel Cargo Logistics, or ICL. Its operations were in New York, in a low-slung complex in the Rosedale section of Queens, on a street dotted by a used-car dealer, an old-school diner, and a bodega. Inside, many of the suites were dark and vacant. The offices of ICL were on the ground floor, down a long, narrow hallway lined with stained tan carpets.

There, in a cubicle inside an eerily silent room, the walls white and devoid of decoration, sat Michael Horan, the ocean manager. He gazed at his computer screen and read a plaintive cry for help from Mississippi.

"I know you're probably hearing this from every client," Walker's email declared. "We're a small business and we received our product shipment this week, however, this container has ALL of our packaging for ALL Christmas products that we intended to sell this year, so we're in a tough position."

He added, "Please let me know if there's anything we can do to expedite the process. And I truly mean we."

Walker was betting on the power of basic courtesy in the midst of what he assumed was a maelstrom of hollering and dysfunction. He figured Horan was besieged by irate shippers channeling their anxiety into issuing threats. He hoped to distinguish himself as a polite person merely asking for a favor. He held out the possibility that Horan might locate a truck going as far as Memphis, or perhaps Dallas. Then he would line up some other option for the very last run.

Horan, an unassuming Queens native, took pride in solving whatever problems came across his desk. Still, he grasped immediately that Walker was asking for something that he might not be able to deliver. Like every other part of the supply chain, trucking was overwhelmed and short of key workers.

"Everything is delayed," Horan told me when I visited his office that fall. "There's a lack of equipment and drivers."

By then, the truck driver shortage had entered everyday American conversation.

It was the subject of congressional hearings and White House task force meetings.

It was a primary reason cited by retail clerks as they told exasperated customers that their sneaker size was unlikely to arrive anytime soon. It was part of the explanation that contractors gave clients as they doubled the time needed to complete their basement renovations.

In the popular understanding, products were scarce and prices were rising because of the ships stuck bobbing off the ports. The vessels were marooned because the docks were full. And that supposedly was because truck drivers had suddenly lost their desire to haul goods.

Given that trucks carried nearly three-fourths of all freight transported in the United States, or more than $10 trillion worth of goods, this was not a minor problem.

Trucking companies were allegedly doing everything they could to recruit and train more drivers, but they could not find enough people. Some blamed it on the extra unemployment benefits the government was dispensing, which—depending on one's politics—were either a necessary dollop of relief amid a public health catastrophe, or a perfidious act of socialism that had sapped the American will to work.

In any event, trucking firms were desperately short of drivers, so consumers were screwed. This assertion took on currency as the crucial fact that made sense of everything else. It insinuated itself into media coverage of the supply chain disruption, repeated as fact by reporters unwittingly parroting industry talking points as they offered explanations for inflation.

The acceptance of this notion was a triumph for the lobbying apparatus that represented the trucking industry, and especially its primary mouthpiece, the awkwardly named American Trucking Associations. For years, it had argued that shortages of drivers posed a mortal threat to American commerce, while demanding public money to train more. The association's chief economist, Robert Costello, had been making that case since at least 2005. The disruptions accompanying the pandemic gave him the ultimate platform for amplifying his pitch.

The trucking industry needed at least eighty thousand more drivers, Costello's operation declared in October 2021, just as Glo's container was unloaded from the *Maersk Emden*. And the shortages were supposedly on track to double to 160,000 over the next decade.

"This is sort of a warning to the entire supply chain, to the motor carriers, to shippers, to everybody," Costello declared at an industry management conference. "The trucking industry alone cannot solve this."

Anytime an industry attests to the impossibility of solving its own problems, this is generally the preamble for a demand for taxpayer-financed subsidies. The trucking industry seized on the shortages of products and rising consumer prices as an opportunity to redouble its pursuit of government handouts for its for-profit training programs—the sort that Marshawn Jackson had endured. And it sought to persuade Congress to loosen the strictures on who was allowed to drive.

The industry was especially keen to lift a federal prohibition that barred people under twenty-one from driving trucks. It offered assurances that entrusting teenagers with rigs weighing as much as eighty thousand pounds and sending them hurtling down the highway, thousands of miles from home, was sure to end happily for all. Here was a ready supply of eager recruits to widen the pool.

The trucking industry needed to perpetuate the idea that drivers were in critically short supply to justify the continued expenditure of public money to train more of them. Yet the permanence of the driver shortage narrative—the fact that the leading trade association had been bemoaning scarcity for decades—gave away the ruse.

It was not that the trucking industry had exhausted the supply of drivers. Rather, it had run out of people willing to endure the misery that driving a truck had become. It could not find takers at rates of pay eroded by deregulation. It was running out of desperadoes willing to sign up for predatory leasing arrangements and other shenanigans employed by trucking firms to force drivers to keep doing their jobs.

In warning of the consequences of the driver shortage, the trade association habitually cited one alarming data point—the driver turnover rate. Among large, long-haul trucking companies, it had averaged nearly 100 percent for decades. This meant that, over the course of the year, the typical trucking firm found itself having to replace nearly all its drivers. A few stayed for years, but most lasted no more than a few months.

The trade association tended to inject this number into the conversation as if discussing an unfortunate accident. In its telling, trucking companies had been doing right by consumers, operating righteously, and treating their people decently, yet, somehow, life had conspired to withhold adequate numbers of drivers. Demographic factors were to blame. Drivers were aging. Young people had other interests. Women simply refused to take part in the secure and wholesome activity of driving across the country by themselves, showering at public facilities, and sleeping wherever—at truck stops, on roadsides, at flophouse motels. This made it impossible for trucking companies to replenish their ranks.

But the trucking industry was not the victim of high turnover rates. It was, in fact, the cause.

Though the trade association accurately noted that many drivers who quit stayed within the industry, simply moving on to better jobs, this was no defense. That so many people opted to switch employers over the course of the year was itself an indictment of how trucking companies operated. People frequently left jobs because their jobs were unbearable.

This insight could be gleaned from academic surveys and expert testimony, and especially from accident data revealing that a truck driver was ten times more likely to be killed on the job than the average American worker. But I did not grasp the foundational truth until I found myself in the passenger seat of a tractor trailer outside a grimy truck stop in the center of Oklahoma as a faint winter sun slid toward the frozen scrub.

The man behind the wheel, Stephen Graves, was exhausted from having driven four hundred miles that day. He needed to use the restroom. He hungered for a decent meal. He was eager to lie down, maybe watch a show, and clear his head. But most of all he required an answer to the same question that dogged him nearly every day as darkness seeped across the land.

Where could he park his rig for the night?

Graves was a different sort of driver than dray operators like Marshawn Jackson. He was known in trucking vernacular as an over-the-road driver, meaning that he typically did not make it home by nightfall. He drove roughly nine thousand miles a month, spending two and three weeks on the road at a time.

By the time he arrived at the truck stop in Oklahoma, he was nearing an eleven-hour federal limit on driving before he was legally required to rest for at least ten hours. This data was captured by a computerized system installed in his rig by his employer. He was inclined to push on for a bit, exhaust his federal time limit, and creep closer to the Texas border. That would shorten the distance to the warehouse next to the Dallas–Fort Worth airport, where he had to drop off his load the following morning. And that would position him to avoid the worst of the rush hour traffic.

But the calculus was tricky. The next stop down the interstate was notoriously short of parking. If he kept going, he might have to pay as much as $20 for a spot, and he might not find one at all. He risked having to settle for the shoulder of a highway on-ramp, passing the night without any services—no shower, nowhere to eat—while worrying about getting robbed.

"You don't want to put yourself in a bind," he said.

This truck stop outside the town of Springer, population 689, was unappealing in the extreme—its bathrooms rank, its dining options limited to shrink-wrapped sandwiches and chicken wings that appeared to have been left in the display case for many hours. There was nothing else around for miles. But it had one thing in its favor: abundant parking.

So Graves pulled into a spot on the cracked pavement and climbed into the bottom bunk of the bed at the back of his cab, a molded plastic enclosure with room for his body, a pillow, and a sleeping bag.

In Tennessee, a guy named Max Farrell had started a company called WorkHound to advise companies on how to retain more drivers. The cause of their problems was straightforward.

"Trucking is seen as a career of second choice," Farrell told me.

WorkHound gave drivers a smartphone app and invited them to use it to lodge anonymous complaints. The feedback it gathered revealed deep-seated frustration over practices that were endemic to the industry.

Drivers were unhappy about being away from their families, and especially incensed by the insensitivity of the dispatchers who dictated their schedules. They felt that their time was being taken from them without compensation by companies that paid them only per mile driven, and not according to the hours they spent on the road. They would show up at warehouses that were themselves short of workers, and wind up waiting there—often for hours—while crews loaded their trailers. They would drive through yards looking for the trailers they were supposed to pick up, their locations often misidentified in the paperwork, and then endure hours waiting for their dispatchers to sort it out.

"People are feeling like they are being used inefficiently," Farrell told me. "A lot of these guys quit because they don't feel respected."

The industry might have responded to this by improving working conditions. It could have paid higher rates per mile along with hourly compensation (something that eventually did happen during the worst of the pandemic, drawing in drivers). But the traditional mode for the largest companies was to perpetually scour the earth for new people willing to give trucking a try and feed them into their training mills.

Central to that strategy was convincing the public and the political system that driver shortages were not only real but a menace to American prosperity; a crisis that justified enormous infusions of public money to train more drivers.

California alone paid out nearly $12 million in tuition to truck driving schools during 2020, which was nearly five times as much as the previous year. Most of that money came from the federal government. And most of the people who graduated from such programs were unlikely to be working as truck drivers even three months after.

The story of the crippling truck driver shortage could not withstand even a moment of scrutiny once the pertinent data entered the conversation. More than 10 million Americans held a commercial driver's license, the credential required to take the wheel of a tractor trailer. That was roughly triple the number of people working as truck drivers. Instead of finding and training new people to give it a whirl, policy should have focused on motivating people who were already qualified to take on the jobs.

Except that would have cost the wrong people money—not the taxpayers, but shareholders of major trucking companies along with their most important customers, big-box retailers that counted on cheap trucking rates.

"This shortage narrative is industry lobbying rhetoric," said Steve Viscelli, the University of Pennsylvania expert. "There is no shortage of truck drivers. These are just really bad jobs."

Much like laboring on the docks and working rail maintenance, driving a truck has always been difficult. Yet until the late 1970s, truck driving had also been among the most lucrative of blue-collar pursuits.

How trucking was downgraded is a story whose structure is by now familiar, a tale of the shareholder seizing primacy through the evisceration of regulations.

In the 1930s, when trucking was still an emerging industry, the business was rife with market manipulation. Trucking firms routinely colluded with one another to keep competitors out and prices high. As part of the New Deal—the federal response to the trauma of the Great Depression—the government imposed a comprehensive regimen of rules.

Under the Motor Carrier Act of 1935, the Interstate Commerce Commission—the same body that had redressed the monopolistic excesses of the railroad barons—gained power to limit market abuses. The law broadly decreed that trucking companies had to offer rates that were "just and reasonable" and barred them from offering special discounts to larger customers.

With a watchdog firmly on the beat, federal regulation limited the number of trucking companies that were allowed to enter the market. It also exempted them from antitrust strictures to encourage them to coordinate their routes and prices, thus ensuring that they could make enough money to keep service reliable. At the same time, the nature of the business—hauling products for retailers—encouraged trucking firms to set up centralized terminals where they could combine smaller shipments into single loads.

The result was an enormous opportunity for organized labor. Truck drivers were working for companies that were, by federal design, consistently profitable. The terminals beckoned as promising targets for union mobilization. Like the ports, they were hubs of commerce that could be shut down to great effect in pursuit of higher pay.

This was the backdrop for the successful organization of truck drivers by an infamous yet effective union, the International Brotherhood of Teamsters.

The history of the Teamsters was inextricably linked to the notorious figure who headed the organization from the mid-1950s until the early 1970s, Jimmy Hoffa. He consorted with organized criminals, spent years in prison on corruption charges, and ultimately disappeared under still-mysterious circumstances in 1975.

But if Hoffa wielded power with brazen disregard for the law, he also illustrated the benefits for workers when they had power to wield.

He captured decisive control over the workforce that kept trucks rolling. Then he struck a cooperative stance with the trucking firms, advancing the most profitable strategies while extracting for workers a hefty share of the rewards as recompense for labor peace.

In 1964, Hoffa successfully negotiated the first national contract for truck drivers, a deal that boosted pay and working conditions for 450,000 people. By the middle of the following decade, 80 percent of truck drivers were represented by a union, and the typical driver was earning nearly $100,000 a year in today's dollars. At unionized trucking firms, drivers were earning half again as much as factory workers.

Those paychecks combined with the militancy of the Teamsters made truck drivers vulnerable to caricature as overpaid impediments to efficiency, much as dockworkers are depicted today. From the halls of Congress to think tanks across the ideological spectrum, trucking became Exhibit A on a lengthy list of industries deemed ripe for deregulation.

Deregulating trucking also emerged as a practical solution to the existential political problem confronting President Jimmy Carter.

Through the late 1970s, the American economy was stuck in the dire straits known as stagflation, a combination of weak economic growth and soaring consumer prices. The country was also contending with an energy crisis exacerbated by the revolution in Iran, a major oil supplier. Carter's economic advisers prescribed deregulation as the effective remedy, a means of injecting dynamism into a moribund economy.

Among economists in that era, the merits of deregulating trucking commanded a virtual consensus. In contrast to rail, trucking was widely thought to be immune to the forces of monopolization. Building a railroad required vast fortunes and years of time, giving existing systems the power to raise rates without risking their businesses. But if a trucking company began lifting prices, fresh competitors could quickly emerge with little more than a few vehicles and drivers.

Regulation itself limited competition by restricting the number of trucking companies allowed to do business. And that gave trucking companies the power to charge higher rates. Those higher rates were like a tax on every consumer, with the benefits absorbed primarily by union drivers. This idea—supported by voluminous academic literature—became the central driving force for deregulation.

In Washington, the only groups opposed to deregulation were the Teamsters and the American Trucking Associations, whose members benefited from less competition. In favor of deregulation was the rest of the business world and grassroots advocacy groups such as Ralph Nader's Consumers Union. Major manufacturers and retailers—including Kraft, General Mills, and Sears—lobbied in favor of deregulating trucking, with the clear understanding that it would bring lower prices.

On July 1, 1980—just three months before he deregulated the railroads—President Carter affixed his signature to the Motor Carrier Act of 1980, making it far easier for new competitors to enter trucking.

"It will remove 45 years of excessive and inflationary govern-ment restrictions and red tape," Carter declared, portraying the new law as a boon for consumers and labor alike. He said the act would "bring the trucking industry into the free enterprise system, where it belongs."

The law worked as designed.

Before deregulation, some eighteen thousand trucking compa-nies had federal approval to haul freight. A year after the act took effect, nearly twenty-nine thousand companies had filed paperwork to launch new trucking businesses or expand existing ones. The re-sult was a sharp reduction in trucking rates.

But so many competitors poured into the business of moving freight by road that few could prosper. By 1985, nearly seven thou-sand trucking companies had disappeared.

At the same time, deregulation dramatically depleted the strength of the Teamsters. By 1997, less than 20 percent of American truckers were represented by a union.

Increased competition, lower profitability, and the extreme dilu-tion of union power all combined to make drivers vulnerable. They accepted longer routes for lower pay. By the year 2000, long-haul drivers were twice as productive while earning only 60 percent of the pay they had brought home prior to deregulation.

They were central components of the Just in Time delivery ma-chine, ferrying a steady flow of product wherever it was needed. Yet they were managed as if their own time was devoid of value. All that mattered was the line item on the spreadsheet showing how many miles they had traveled.

By 2018, median wages for truck drivers had dropped by more than one-fifth since deregulation, and by half in some parts of the country. Marshawn Jackson's experience had become typical. Peo-ple were entering training programs run by major trucking compa-nies and agreeing to financial terms that obligated them to continue driving for those firms for months and even years—long after they learned that they would never earn the pay they had been promised.

As one expert put it, long-haul trucks had become "sweatshops on wheels."

Stephen Graves vehemently rejected such characterizations.

As I rode along with him in his tractor trailer in January 2022, traversing the frozen midsection of the United States, he rhapsodized about the open road. He found poetic beauty from the cab of his giant rig. The flat and treeless plains unfolded before him, covered in snow and punctuated by distant herds of cattle. He anticipated his favorite stretches of highway for hundreds of miles.

"I just love this area," he told me, as we rumbled through the Flint Hills of Kansas. "In the summer, the tall grass, the prairie grass is going full. It blows gently in the wind. You can just listen to the wind. It's such a calm and soothing feeling, maybe like one of those sounds that you can buy and listen to for meditation or relaxing or sleeping. It's very therapeutic."

Graves was about to turn sixty-five. He had worked as a long-haul truck driver for more than two decades, and he took pride in the comforts financed by his miles on the road. He owned a condo in Tennessee, where he liked to putter around repairing old shortwave radios or sit on the couch reading cold war history books. He was a world traveler who was eager to share the details of his journeys—to England, Australia, Russia.

He felt a measure of satisfaction that he was engaged in something useful.

"I don't pretend that I'm Superman," he said. "I'm a cog in the wheel. If I don't do what I do, things will slow down. Somebody bought a new TV. They're waiting for it. They're going to watch the game."

Graves allowed that trucking was not for everyone, but it provided a reliable, more than adequate standard of living if you did it right, he said. He held no debt to his employer, American Central Transport, a company based in Kansas City. He was free to move somewhere else anytime he wanted.

Still, the three days I spent in his passenger seat, and the two nights I slept in his upper bunk, provided a glimpse into a way of life whose challenges were exceptional by any ordinary reckoning.

Here was a job that entailed physical exhaustion and perpetual anxiety—over the availability of parking, and the need to balance caffeine intake against the logistical inconvenience of having to stop for a restroom. Here was a life in which a hamburger consumed inside a fast-food outlet constituted a sit-down meal, as opposed to a microwave burrito wolfed down on the interstate.

"The lifestyle probably is the first thing that smacks people in the face," Graves said. "You know what it does to you. You're thinking about it all the time. We're tired. Our bodies are starting to go. Our bladders have been put to the test. And no exercise. We end up with all types of heart and other health ailments. We have drivers that have diabetes, that have heart issues, weight issues."

The public health literature confirmed that long-haul trucking tended to decimate sleep and increase blood pressure and sugar levels, while yielding substantially greater risk for heart disease and diabetes.

Graves was never free of the awareness that any momentary loss of focus, any minute error by some other driver distracted by their phone or a crying baby or an itch, could result in catastrophe.

Only seven months into his career, he had been carrying a load of electronics from North Carolina to Virginia, traveling north on I-95, when a pink Cadillac Escalade moving in the opposite direction hurtled over the divider, flipped in the air, and landed 150 feet in front of him. He swerved but still clipped the vehicle. He was certain that everyone inside the other vehicle was dead. Somehow, the couple had survived.

"If this isn't scary, you're a fool," Graves said. "It takes more than the length of a football field to stop out here."

I had been paired with Graves by American Central Transport. The company was clearly one of the better operations in the trucking business. It was a midsize firm that ran a fleet of three hundred trucks, hauling loads between Minnesota and Texas, and as far east as the Carolinas and Georgia. It had increased pay twice during the course of the pandemic in a bid to hang on to its drivers and recruit new ones. Its turnover rate was a "mere" 64 percent—an insane number in nearly any other industry, yet far better than average for trucking. Its most experienced drivers were earning more than $90,000 a year. The restrooms and showers at its yard in Kansas City were clean and comfortable.

I had asked for a ride-along so I could see for myself what truck drivers experienced out on the road. That the company was willing to accommodate me attested to its confidence in the virtues of its enterprise.

Still, management was not going to entrust the passage of a nosy journalist to an unproven newbie or malcontent. Graves was clearly a pro's pro, and a guy who professed to enjoy his job. I was seeing the long-haul journey in the best possible light. This was as good as trucking got.

My three-day ride-along brought home why trucking firms were then dispensing $10,000 signing bonuses in frantic pursuit of available drivers; why recruiting agents were approaching drivers at truck stops, proffering higher pay in trying to entice them to jump ship; and why every logistics expert assumed that the driver shortage was a trend that would persist.

Graves had envisioned a different life.

Raised an only child near Richmond, Virginia, he attended West Virginia University with dreams of being an engineer. But when his father became ill, he dropped out of college to help his mother.

He worked on a state road-building crew and as a quality control technician at a Coca-Cola bottling plant. He installed equipment for a telecommunications company. When he lost that job, he went to the unemployment office for an assessment of possible career paths.

He considered training to be a plumber or carpenter, but those professions entailed years of apprenticeships. He settled on trucking because it required only a few weeks of school before he could be out on the road earning.

Some 1.6 million miles later, he was still behind the wheel.

I joined him just after New Year's in 2022, on what was day ten of a nineteen-day trip that had already led him across Arkansas and into Texas, through Chicago on three separate loops, into Indianapolis, and Spartanburg, South Carolina, before bringing him to the company yard in Kansas City.

His day had begun at 3:30 that morning. Graves had emerged from his bunk at the rear of his blue Kenworth T680. He slipped on a Day-Glo orange woolen hat, opened the cabin door, and climbed two ladder-like steps down to the pavement, grimacing as he landed—he suffered flat feet—while eyeing the yard suspiciously. The asphalt was pockmarked by patches of black ice and crusted snow.

In the twelve-degree chill, he applied a metal gauge to his tires to check the pressure. He examined his brake lines. He checked his brake pads and windshield wiper fluid, ensuring that they had not frozen. He scrutinized the connection between tractor and trailer.

Satisfied, he returned to the warmth of his cab, fortified with coffee, and then checked the paperwork on the load he was to pick up that morning. He released his brakes and rolled out of the yard.

He was headed to a warehouse thirty-five miles southwest of Kansas City, where he would drop the empty trailer he was pulling, and then pick up a full one loaded with twenty-six crates of tractor parts. He would haul that cargo 545 miles south, across the plains of Kansas and Oklahoma, to a distribution center in Fort Worth, Texas.

Graves kept his cabin temperature cool, at sixty-three degrees, to maintain alertness and stave off highway hypnotism.

Every truck stop was like a shrine laid out to ward off the demons of fatigue. Refrigerated display cabinets were stocked with supercaffeinated energy drinks whose brand names attested to the swapping of health for the pursuit of an immediate jolt. *Red Bull. Java Monster. Bang.*

"It's a deal with the devil," Graves said. He stuck with coffee.

At the warehouse, he consulted the computer mounted on his dash for precise instructions. They directed him to Building 2, which Graves logically assumed meant the second building that he encountered. But after climbing out of his cab and limping up a staircase, he was directed back to the first building.

This stop was known in trucker parlance as a "drop and hook"—every driver's favorite kind of pickup. He merely had to back into the assigned space, detach his trailer and leave it behind, and then hook his tractor to the next load. The alternative was waiting around for an indeterminate period while workers placed the cargo into his trailer.

The accompanying paperwork showed him delivering his cargo not to Texas but to Ohio. He sat patiently in a folding chair before a clerk set things in order. This, Graves handled with studious cheer.

"Good morning, ma'am," he said to every woman he encountered—at truck stop cash registers, at warehouse reception areas. "Are you doin' okay?"

As the miles slipped past and the landscape shifted, the gnawing loneliness of the road was the single unchanging feature of his reality. He seemed to savor the most rudimentary of human connections.

He made a point of learning the name of the woman who poured him a coffee at a Burger King in Oklahoma—Bailey. He talked about the kindness in her glance for hundreds of miles after.

"I try to give everyone a smile," Graves said. He was actively compensating for the others on the road. "Drivers are generally snarly, because they are tired, they're hungry, and their schedules suck, and they tend to take it out on other people."

We were moving southwest down I-35, crossing from Missouri into Kansas. We rolled past an assisted-living facility, a Harley-Davidson motorcycle dealership, an Applebee's restaurant, and strip malls full of nail salons and check cashing places. We passed a Hostess Twinkie factory and an indoor skydiving place.

Graves clearly reveled in challenging popular stereotypes of the truck driver. He spoke with the precise elocution of a college professor. He rejected the term "truck driver" for "commercial driver." And he scoffed when I asked him if he used a CB radio. He had discarded it years ago, weary of hearing crude and hateful chatter dispensed over the crackly airwaves.

He tended to begin his mornings listening to the global news report from the BBC World Service in London, before switching to light jazz or classical.

"I love Brahms," he told me as we wound through Kansas.

Above all, he celebrated his constant motion as freedom from the cubicle life that confined many workers. He was seeing the country. He was taking in the majesty of the land.

But as he hewed almost exclusively to the interstates, what he mostly saw was the dreary service corridors of American life, a generic blur of gas stations, fast-food outlets, chain motels, and liquor stores.

"After a while, all the cities and towns run together," he said. "I'm seeing the world through a windshield."

Driving a truck was especially difficult for younger people, Graves told me. He sometimes spoke to classes of fresh trainees, where he dispensed a regular piece of advice: maintain $500 in cash at all times for a plane ticket home.

One of the primary reasons most did not last, he explained, was the challenge of retaining connections to the rest of the planet.

Wives, girlfriends, and children constantly wanted to know when they were likely to be home—often an unanswerable question. Relationships frequently did not endure the trials of the road.

Graves had long made peace with his own solitary existence.

More than a decade earlier, he had met a woman in Tennessee at a restaurant. They went out on a few dates. Then, she began asking him how long he planned to continue driving.

"She said, 'I'd like for us to have a relationship,'" Graves recounted. "I was flattered, but you know, what am I going to do if I just stop working? I have no income. I have no job for the time being. I just have love. That's nice for a couple of days but, you know, love is not automatically deposited in my payroll."

The skies were a sullen gray as we passed through the monotonous sprawl of Oklahoma City, just after four in the afternoon. The interstate widened to three lanes.

Under federal regulations, Graves was required to take a thirty-minute break within eight hours of driving. An electronic monitoring device installed in his cab showed the seconds ticking away, with less than ninety minutes left. His body was stiff. He could have used a stretch. But he did not want to stop in Oklahoma City, not with rush hour building. He pressed on, past a Cracker Barrel, a Taco Bell, a Super 8 motel, a high school football stadium.

By the time he reached the truck stop south of Springer, the pale sun was grazing the horizon.

In trucker lingo, any gas station with food and parking was known as an oasis. This lot was full of trucks parked tightly next to one another on the cracked pavement, their engines idling to produce heat.

Fort Worth was more than one hundred miles to the south. But the fear of not finding parking further down the road won out. Graves backed into a space between two other tractor trailers and headed into the shop in search of sustenance.

He selected a turkey sandwich, plus two Pop Tarts—"one for dessert, and one for breakfast." He walked back to the truck in the darkness, a crescent moon overhead. He ate his dinner and slid into the lower bunk.

Early the next morning, he pulled out of the lot. Just before the Texas border, the skies still dark, he drove past an enormous Wynn's casino, the facades decked out like world landmarks—the Colosseum in Rome, Buckingham Palace, the Chrysler Building.

In Fort Worth, he navigated a tangle of cloverleaf merges and located his destination in a warren of warehouses. He dropped off his load and then continued south, as he listened to a report about the Consumer Electronics Show in Las Vegas on the BBC.

"People buy too much crap," he said.

Fifteen miles down the interstate in Grand Prairie, Texas, he hitched his tractor to a trailer bound for a Walmart distribution center near Kansas City. The paperwork showed that he was picking up thirty-eight thousand pounds of contact lenses, dog food, salsa, and ground coffee.

All day, he retraced his route into Oklahoma up I-35. He passed the night at a truck stop outside the town of Tonkawa. The next morning, he was on the road by five, crossing into Kansas as a fiery sunrise seeped from the frozen plains.

He dropped his trailer at an enormous Walmart lot just after 10 a.m. He was on track to reach Kansas City by midday, with a precious afternoon off. He planned to leave his tractor in the company yard and shell out more than $100 for a night at a nearby hotel. He was already savoring the hot shower, the afternoon nap, the maid service.

"It's always nice to have somebody come in and make up your bed, leave you an exotic swan made out of towels," he said.

But an hour later his terminal beeped, and a computerized voice alerted him to an incoming message from a dispatcher in Kansas City. Another truck carrying pet food had broken down near Columbia, Missouri. Could he divert there—two hours away—to rescue the load and carry it to a distribution center in Joplin, Missouri?

The message was missing several pertinent details—times, addresses, phone numbers. With evident irritation, Graves dialed the dispatcher, only to discover that he was on his lunch break. He called another dispatcher.

"Let's talk about this cryptic message you left me," he said, clearly laboring to maintain restraint.

"That's the one thing I hate about office people," he told me as he sat on hold. "Why send me a message? Just *call* me."

He rerouted to Columbia, trading in his hotel bed for another night in his bunk at a truck stop.

"I'm tired, man," he said. "I kind of want to stretch out. But, hey, I'm just a machine, right?"

Despite the long odds, Michael Horan managed to rescue Glo's container from its holding point in Southern California.

"Hi, Hagan," he replied to Walker's plea for highway transport. "I have a truck picking up tomorrow from Los Angeles and will deliver to you on Tuesday." He was talking about November 9, leaving plenty of time to manage holiday shipments. He could arrange this for an additional $1,100.

"Yes!" Walker wrote back immediately. "Let's do it!"

Just before eight in the morning on the appointed Tuesday, Walker drove to his warehouse, an abandoned furniture showroom on the ragged edges of Starkville.

The tractor trailer was already there, backed up to the loading dock. The driver, who was based in Southern California, was resting inside the cab after his four-day, 1,900-mile transcontinental journey.

As the driver lifted his rear gate, Walker peered in excitedly. There sat 1,595 cartons stacked atop twenty-four wooden pallets. He and three of his employees wielded electric pallet jacks to move the goods into the warehouse.

An hour later, a Glo truck began collecting boxes of the newly arrived cargo, hauling it to the old movie theater downtown.

Inside, Walker's employees began preparing shipments for customers across the United States and around the world, as far away as Singapore.

Walker felt an unfamiliar sense of relief. Through pluck, determination, and no small measure of luck, he had managed to navigate the pitfalls of the Great Supply Chain Disruption.

Elmo had completed his trip in time for the holidays.

CHAPTER 16

"THANK YOU FOR WHAT YOU'RE DOING TO KEEP THOSE GROCERY STORE SHELVES STOCKED."

How the Meat Industry Sacrificed Workers for Profits

As Elmo finally arrived at Glo's warehouse in Mississippi in the middle of November 2021, elsewhere in the United States, shortages of goods persisted and even worsened.

Inflation on a range of consumer items was soaring. Some of this was the result of transportation bottlenecks. But a lot of the upheaval reflected forces that long predated the pandemic. Decades of deregulation had deactivated antitrust law as an impediment to a future ruled by scale and efficiency. In many industries, businesses had succeeded in amassing monopoly power on a scale that would have impressed the Robber Barons.

This is a story that is central to understanding the breakdown in the supply chain. The shortages were not an accident, but rather the outgrowth of a concerted strategy. For decades, some of the largest businesses had amassed chokeholds on their markets while limiting the supply of their products as a way to charge higher prices.

Even as the pandemic severely disrupted their access to materials, they were perfectly positioned to reap gains. They seized on scarcity, fear, and tragedy as an opportunity to profiteer, racking up record profits as they lifted prices. Nowhere was this reality more discernible than in an enterprise central to the American story—the meat industry.

San Twin begged her mother to stay home.

It was March 2020, and COVID-19 was spreading inside the slaughterhouse where Tin Aye worked in Greeley, Colorado.

San Twin did not understand the contours of the danger, or how this lethal virus spread. But it seemed safe to assume that it was being accelerated by the very conditions in which Aye labored.

Inside the vast, dank slaughterhouse, as many as 1,500 people per shift carved cattle into cuts of beef, processing more than 1 million animals per year. They stood for hours on an assembly line in close proximity to one another, amid spraying water jets and sanitizing chemicals misting down from the ceilings. Some wielded knives and saws to hack away at sides of beef, sending blood spurting. Others scooped away innards and excrement.

Aye's job was plucking finished cuts of meat from the assembly line, packaging them in plastic, and putting them in boxes. She and her coworkers sweated and breathed heavily as they attended to their duties, straining from exertion. Some wore face masks they brought to work themselves, but this tended to fog up safety glasses, so many went without.

A man who stood just behind Aye on the plant floor had already contracted COVID-19. Workers nervously speculated about who would be next.

The slaughterhouse was owned by JBS Foods, the largest meat

processing company on earth. A Brazilian conglomerate, the company had entered the American market in 2007 with the $1.4 billion purchase of Swift. That deal had included the Greeley plant. Since then, JBS had broadened its reach to encompass one-fourth of the American capacity to slaughter beef.

The two brothers who oversaw JBS, Wesley and Joesley Batista, carried a reputation for hard-charging business practices that verged into criminality. Back in Brazil, an explosive scandal had revealed how they had bribed three presidential administrations and a key official at the Brazilian Development Bank in a scheme that secured nearly $5 billion for their international expansion spree. Much of the cash for their purchase of Swift had come via that endeavor.

Both brothers had spent about six months in prison in Brazil. By the time the pandemic arrived, they had been out for two years, their meat empire intact, with a collective fortune worth more than $5 billion.

Their business was perfectly positioned to exploit the panic of the pandemic, as families stuck at home compensated for the shutdown of schools and restaurants with home-cooked meals and heftier portions.

But cashing in required that workers like Tin Aye continued to show up at slaughterhouses.

As fear among the workers in Greeley mounted, JBS remained tight-lipped about how many people inside the plant had contracted the coronavirus. By the end of the month, more than eight hundred plant workers were refusing to turn up for shifts, fearful of bringing COVID-19 home to their families. Workers at other slaughterhouses across the country were staying home, too.

San Twin was tormented by the thought that her sixty-year-old mother was spending every workday inside a densely packed COVID-19 hotspot.

A member of the Karen people, an oppressed ethnic minority in Myanmar, Aye had survived a harrowing escape from her homeland while pregnant with her only child. Together, they had sustained themselves for fifteen years in a refugee camp inside Thailand, living in a bamboo hut without electricity or plumbing. They had subsisted on rice and beans donated by the United Nations, supplemented by whatever cash Aye could earn cleaning houses, doing laundry, and tending pigs.

They were eventually offered a choice of countries in which to settle, among them Canada, Australia, and Norway. They picked the United States, having heard that this was a place where a hard-working immigrant could always find a job.

They landed in Denver in the summer of 2012, knowing no one and speaking no English. Aye got a job working nights at the slaughterhouse in Greeley, an hour's drive north. She carpooled with other Karen immigrants, leaving for the plant in the early afternoon and returning at four in the morning.

The work was grueling and relentless. Aye came home with an aching back, swollen fingers, and bruises on her legs. She also returned soaked—not only because of the misting chemicals and spray jets, but because bathroom breaks were so infrequent that she sometimes urinated in her clothes.

She started at $12 an hour, which seemed like a lot of money.

As the pandemic emerged, San Twin's fear of losing her mother was compounded by the fact that she was pregnant.

"My mother was the only family that I had," Twin told me as we sat on the floor of her home in Denver in December 2021. She was nearly thirty, and about to open her own restaurant.

Her son Felix, then twenty months old, sat quietly in her lap, gazing at the older woman in the framed photo on the living room wall. Aye wore glasses and was draped in a white lace shawl, her steely resolve giving way to a warm smile. Here was the grandmother that Felix would never meet. Twin's pleading had failed to persuade her mother to stay home.

"I said, 'Please don't work in the plant anymore,'" Twin told me. "She said, 'I have to pay the bills. I'm strong. I'll be okay.'"

Like the railroad workers and millions of other mainstays of the American supply chain, Aye and the other employees inside the slaughterhouse lacked paid sick leave. They, too, confronted a grim choice between their lives and their livelihoods.

Aye was among the nearly three hundred workers at the Greeley slaughterhouse who were hospitalized with severe COVID-19 symptoms during the first four months of the pandemic, and one of at least five people who died. And the Greeley plant was merely one facility within an enormous industry consumed by the virus. Across the United States, 59,000 meatpacking workers contracted COVID-19 over the course of 2020, with 269 people dying.

That so many ordinary workers found themselves in harm's way in the midst of a pandemic was no mere misfortune. It was a direct outgrowth of the business plans pursued by their corporate bosses. Just like the railroads and the trucking companies, the largest agribusiness conglomerates had placed the imperatives of their shareholders above the welfare of their employees.

Worse, they manufactured fears of a meat shortage to gain the complicity of the American government. They fomented public alarm over potential disruption to the food supply as a way to justify the ultimate sacrifice of their workers—all in the service of boosting their profits.

On April 18, 2020, a doctor at a hospital in the Texas panhandle emailed JBS to warn the company that its slaughterhouse in the nearby town of Cactus was the source of an intensifying wave of COVID-19.

"One hundred percent of all COVID-19 patients we have in the hospital are either direct employees or family members of your employees," the doctor wrote. "We believe there is a major outbreak of COVID-19 infection in your Cactus facility."

He added: "Your employees will get sick and may die if this factory continues to be open."

But the plant remained in operation. Ten days later, President

Trump intervened to ensure that it would stay that way. He cited the Defense Production Act—a law dating back to the Korean War—in signing an executive order barring the shutdown of meatpacking plants as a threat to national security.

"Such closures threaten the continued functioning of the national meat and poultry supply chain, undermining critical infrastructure during the national emergency," Trump's order declared. "Given the high volume of meat and poultry processed by many facilities, any unnecessary closures can quickly have a large effect on the food supply chain. For example, closure of a single large beef processing facility can result in the loss of over 10 million individual servings of beef in a single day."

In essence, Aye had to continue risking her life, or Americans would risk going hungry.

This formulation from the Trump administration constituted a decisive victory for the meat industry. It validated a propaganda assault aimed at stoking fear of food shortages as the means of maintaining business as usual.

As internal emails unearthed by *ProPublica* later revealed, Trump's order echoed key talking points crafted by officials at the North American Meat Institute, the grandly named advocacy shop for the industry. This was not a coincidence. Top agribusiness executives had been meeting and corresponding regularly with senior Trump administration officials to plot strategy engineered to keep slaughterhouses open.

The most critical element to this partnership between the White House and the meat industry was their dissemination of the false theory that closing slaughterhouses jeopardized the nation's access to food. This was a point that key industry executives had been voicing loudly to ratchet up anxiety.

Two weeks before Trump's executive order, the president and chief executive officer of the meat conglomerate Smithfield Foods, Ken Sullivan, raised the specter of empty grocery store shelves as he announced the closure of a pork processing plant in South Dakota.

Such disruptions were "pushing our country perilously close to the edge in terms of our meat supply," Sullivan said in a press release. "It is impossible to keep our grocery stores stocked if our plants are not running."

Yet at the same moment that Smithfield issued those frightening words, American meatpackers were collectively sitting on 622 million pounds of frozen pork—far more than they held before the pandemic. The largest meatpackers in the United States had enough inventory to stock the shelves of every grocery store in the land. They were in control of an empire so vast that they were increasingly selling their products to households on the other side of the world. Over the course of 2020, Smithfield and JBS both dramatically expanded their exports of pork to China. Their supposed concern for the sanctity of the American food supply did not preclude them from sending home-raised meat across the Pacific.

The JBS plant where Tin Aye worked was one of the nine largest meatpacking plants in the country. Typically, it exported nearly one-third of its output to twenty different countries. The notion that she had risked her life to feed the nation was both tragic and ludicrous.

Smithfield's warning of meat shortages was such a brazen lie that it prompted ridicule even within the National Meat Institute.

"Smithfield has whipped everyone into a frenzy," wrote the head of communications at the trade association, Sarah Little, in an internal email obtained by a congressional investigative panel. She added that Sullivan, the CEO, was "directing the panic," even as he was focused on promoting exports.

"So basically the meat shortage story is that there is no shortage," wrote another member of the trade association staff.

But the most important audience for the tale of vanishing grocery stocks, the Trump White House, not only bought the deception but acted on it.

In the middle of March 2020—just as San Twin was imploring her mother to stay home—representatives for major meatpacking companies began meeting with senior officials at the US Department of Agriculture, seeking action that would compel slaughterhouse employees to continue showing up.

Trump's Department of Agriculture escalated the warnings of a threat to the food supply, passing them on to Vice President Mike Pence. The campaign produced a recommendation from the Department of Homeland Security that states should classify slaughterhouse workers as "critical infrastructure." That designation was aimed at ensuring that workers would continue turning up for shifts even in communities where local quarantine rules and other social distancing measures called for them to stay home.

At the same time, executives for the largest meatpackers, including JBS, personally beseeched Trump's secretary of agriculture, Sonny Perdue, for help in persuading slaughterhouse workers that showing up for their shifts was their best option. They specifically discussed how to convey to workers that they would not qualify for government benefits if they stayed home.

On April 3, 2020, the executives held a call with Perdue to ask that he press the White House to involve Pence or Trump in their undertaking. Following that call, they sent an email to Perdue thanking him for his time and urging him to orchestrate "a strong and consistent message from the President or Vice President" that "being afraid of COVID-19 is not a reason to quit your job and you are not eligible for unemployment compensation if you do."

Four days later, Vice President Pence gratified the industry during a Coronavirus Task Force briefing at the White House. He expressed concern over "incidents of worker absenteeism" at some slaughterhouses, which had led to "some plants having reduced capacity." He urged slaughterhouse workers to keep the assembly lines moving.

"Thank you for what you're doing to keep those grocery store shelves stocked," Pence said. "You are vital. You are giving a great service to the people of the United States of America. And we need you to continue as part of what we call our critical infrastructure to show up and do your job."

By the time Pence delivered those words, Tin Aye was in the intensive care unit at a Denver hospital.

She had labored for weeks through severe coughing and a high fever while her daughter had implored her to go to the hospital. Only when she was struggling to breathe had she finally sought medical attention, setting aside her reluctance to spend the money for care.

That same week, San Twin was in another hospital, having delivered Felix via an emergency caesarean section after she suffered painful contractions and shortness of breath. A test revealed that she, too, had COVID-19.

The day after Felix was born, Aye called from her own hospital bed to deliver the news that, according to her doctors, her COVID-19 was extremely advanced.

"She was calling to say goodbye," Twin told me. "She said, 'I really want to see you, but I can't see you anymore.' She told me to work hard for Felix. Just believe in the positive view and help yourself and others. And then she dropped the phone, and I never talked to her again."

Aye suffered two strokes and slipped into a coma. She was kept alive by a ventilator until she drew her final breath on May 17, 2020.

As I spoke with Twin more than a year later, she could not shake the sense that her mother had been used and discarded by a rapacious industry. Her gut-level sentiment took affirmation from the astonishing profitability of the enterprise in which Aye had labored.

During the first two years of the pandemic, the four largest meatpackers showered shareholders with more than $3 billion in dividends. The year that Aye died, JBS celebrated revenues of $22 billion on sales of beef in the United States.

"The results for 2020 make us very proud," the CEO of JBS, Gilberto Tomazoni, told investors.

The company gave Twin $6,000 to help with her mother's funeral arrangements and never called to offer condolences, Twin told me.

The Department of Labor's Occupational Safety and Health Administration later cited JBS Foods for "failing to protect employees from exposure to the coronavirus."

The federal body levied a fine of $15,615. That was less than what JBS earned on its sales of beef in the United States every thirty seconds.

"They did not care about human beings," Twin said. "They only care about the money."

Though the meat shortage narrative was cynically concocted, it was effective precisely because it had a veneer of truth to it. Some grocery stores were running short of beef. But this was not a sign of a crisis befalling the industry in the face of trouble. Rather, the system was running according to the designs of the executives who constructed it. They managed the supply toward keeping it perpetually tight, because less meat on the market allowed them to demand higher prices.

The American beef supply was dominated by four enormous companies—JBS, Cargill, Tyson Foods, and National Beef. They operated huge and centralized slaughterhouses that processed cattle trucked in from vast swaths of territory, collectively yielding 85 percent of the steaks, roasts, and other cuts of beef consumed every year. Anytime something happened to disrupt production at any one plant, the consequences were swift and palpable.

The meatpackers were not bystanders to this reality. They were the orchestrators and beneficiaries. Over the decades, they had rigged the system in their favor. They used their dominance to ensure that the independent ranchers who raised cattle had few places to sell them, limiting the prices they had to pay for the animals coming in their slaughterhouse doors. At the other end of their operations, in their sales to supermarket chains, wholesale distributors, and restaurants, they wielded scale to limit the supply of steaks and ground beef, boosting the prices of their products. Here was the logic of modern-day profit maximization applied to the old-school domain of red meat. The companies in control of the meat supply had reclaimed for themselves the market power of their forebears, the Robber Barons.

The four giant companies had exploited the ardor for deregulation that had shaped every presidential administration for half a century. They had unleashed a wave of mergers that had concentrated the market to extreme levels, and then reduced capacity by shuttering slaughterhouses.

Fewer places to kill and process livestock left cattle ranchers facing limited options for markets to sell their animals. That forced ranchers to take whatever price they could fetch from a smaller universe of buyers. And that limited the price that meatpackers had to pay for cattle, boosting their bottom line and providing fatter dividends for shareholders.

In an age ruled by the imperatives of lean production and profit maximization, engineering scarcity was a proven recipe for success. But it came at the expense of consumers along with the people who labored to produce meat.

Much as consolidation in the railroad business had proved rich for investors and harmful to shippers and rail workers, the domination of the meat industry by a handful of players had lifted prices for food while cutting ranchers out of their traditional share of the gains. The pandemic was not responsible for this state of affairs, but it exposed it, yielding a shock that the meatpackers were perfectly positioned to exploit. Here was a fresh opportunity to limit production while increasing prices for seemingly legitimate reasons.

Here, too, was the latest validation for a time-honored commercial strategy: the wielding of monopoly power.

CHAPTER 17

"WE DO NOT HAVE A FREE MARKET."

How Monopolists Exploited the Pandemic

"You could see a cow across the road," said Annika Charter-Williams, a fourth-generation rancher. "And you couldn't find ground beef in Billings, Montana."

Hagan Walker's worst brush with dysfunction had come on the water. For the Charter family, trouble played out across the land. Even as Americans were sequestered in their homes, taking solace in backyard barbecues, the Charters' ranch on the high plains of Montana struggled to find takers for its cattle. Yet at chain super-markets in the nearest town, many meat display cases were empty.

By every indication, it should have been an especially rewarding time to be a cattle rancher. Americans were taking refuge in food, consuming record volumes of beef. They were paying prices that had increased by more than one-fourth over the previous year, with some cuts of meat soaring by 70 percent.

But somewhere between American dinner plates and the Charter family's eight-thousand-acre ranch, their share of the proceeds from the $66 billion enterprise of raising beef in the United States had gone missing.

Steve Charter, Annika's father, had the calloused hands and tattered coveralls of a man who spent the entirety of his life on the ranch. He had begun driving an antique tractor when he was only eight, using it to sweep freshly cut hay into piles for baling. He was accustomed to working seven days a week, 365 days a year, in winter temperatures descending to minus forty, and in summer swelter reaching 110.

On the morning I met him in December 2021, we rumbled up a snow-crusted road in his feed truck, hauling a mixture of grains to his herd of mother cows and calves. As we approached, the plaintive mooing and groaning of two hundred animals swelled into an insistent chorus.

"They'll be much happier once we feed them," he said.

Charter's two grown children—Annika, then in her midthirties, and her brother, Ressa, forty—had both built homes on the ranch, living there with their own families. They were passionate participants in the ranching life.

"It's in our blood," Ressa told me. "It's all I think about."

The highlight of their year was an annual two-day trail ride from the ranch to the summer range. The entire family mounted horses and guided the herd up into the verdant grasslands of the Bull Mountains for a season of grazing.

The Charters were active in a local environmental organization, the Northern Plains Resource Council, which promoted soil conservation by working to limit the size of herds and plant appropriate crops. The central goal was to mimic the role that the bison had played in centuries past as they roamed across the unbounded continent, spurring the regeneration of native grasses without overgrazing individual areas.

"Cows can be very destructive environmentally, or they can be the greatest tool that we have," Charter told me. "It takes generations to learn this."

Then nearing seventy, he had long envisioned his six grandchildren staying on the ranch and continuing the family's way of life. But that prospect was being steadily eroded by the monopolistic incursions of the meatpackers.

Charter was what was known in the business as a cow-calf producer. He held on to a core group of two dozen heifers, introducing bulls to sire calves that were typically born in June. By the time the calves reached eight or nine months old, he faced a critical decision. If the price for live cattle appeared promising, he was inclined to keep his calves, bearing the costs of their continued upkeep, before selling them the following year at a premium.

But if prices looked weak, he tended to unload his calves earlier. He generally sold to so-called feedlots that put them on a nutrient-rich diet of corn and other grains engineered to add bulk. The feedlots were typically affiliated with one of the four major meatpackers. They would eventually sell their cattle to one of their slaughterhouses, such as the JBS plant in Greeley.

Prices for Charter's cattle had been chronically weak for years. Every spring, there seemed to be fewer buyers for his animals, forcing him to take any price offered by the conglomerates that controlled the slaughterhouses. Every cycle, he lost a little more faith in the endurance of his way of life.

The ranch had not been profitable for five years. That fact was forcing Charter to confront the once unthinkable fate that had befallen many of his neighbors, along with more than half a million other American ranchers in recent decades: selling off his herd, and perhaps handing the land to a developer, turning another working Western expanse into an exclusive getaway for a tech billionaire or master of finance.

"We are contemplating getting out," Charter told me, his voice catching as he choked back tears. "Basically, all my life, our percentage that we get of the consumer dollar has just been going down. In the last couple of years, it's just gotten to the point of ridiculousness."

This would have been easier to accept had other ranchers been prospering.

"If I'm not a good producer, the market should be able to tell me that and say, 'Well, you can't be in business, because you can't produce economically,'" Charter said. "And I will live and die by that. But we do not have a free market, and it doesn't give a reward to the right people."

Many of the shortages resulting from the disruptions of the pandemic could be boiled down to some version of production and logistics challenges. The factories that made electronics in China had been forced to halt operations, so these products were scarce. Containers full of goods sat floating off ports, so retail shelves were empty. Truck drivers were less willing to perform a miserable job, so companies found themselves waiting for deliveries.

All of these troubles were enhanced by the established supremacy of Just in Time and *lean* inventories, which had eliminated the margin for error. This explained why Walker had faced such difficulties in securing the materials needed to make his order.

But that was only part of the story. The same forces at work in the breakdown of the American railroad system—monopolists reducing capacity—were operative in huge swaths of the economy.

Companies that produced and distributed a vast range of goods, from shaving cream to medicine, had captured dominant holds over their markets while concentrating production in enormous, centralized factories. That allowed them to limit supply, which tended to push prices higher. And it left consumers vulnerable to shortages and price spikes whenever something went awry.

The journalist Barry Lynn had once dissected the impact of a single merger, the 2005 takeover of Gillette by the household goods behemoth Procter & Gamble. Through a variety of brands that conveyed the illusion of robust consumer choice, the combined entity sold more than 75 percent of men's razors, and roughly 60 percent of laundry and dishwasher detergent. It sold half of all toothbrushes, batteries, and feminine pads, and two-fifths of all toothpaste, diapers, and nonprescription heartburn drugs. One single brand—Bounty—controlled nearly 45 percent of the market for paper towels.

Lynn detailed how Kraft, then the world's second-largest producer of food, had used a 2000 merger with Nabisco to capture major brands like Oreo cookies and Planters peanuts. Then, Kraft shut down thirty-nine processing plants and eliminated one-fourth of its products to boost its profit margins.

Engineered scarcity was an excellent way to lift prices and satisfy shareholders, but it came at the expense of reliability. Every shock risked producing shortages. The consequences frequently involved more than ersatz chocolate cream cookies.

In the spring of 2022, parents in much of the United States found that baby formula had suddenly become nearly impossible to find, an emergency for many families. The immediate explanation seemed reasonable. A plant that made formula in Michigan had become contaminated with a lethal strain of bacteria, an episode blamed for the deaths of two babies. The operator of the plant, Abbott Labs, had shut the factory and issued a recall, yielding shortages.

But as the monopoly expert Matt Stoller pointed out, Abbott was not some niche supplier of baby formula. It controlled 43 percent of the market, selling through popular brands such as Similac. Abbott and two other major conglomerates, Mead Johnson and Nestlé, collectively owned 98 percent of the market.

In a truly competitive market, companies would feel pressure to safeguard the quality of their products and ensure adequate supplies, or otherwise risk losing sales to rivals. But dominance meant that rivals were few, diminishing the usual incentives for responsible stewardship.

Abbott was a case in point. Seven months before the outbreak, a whistleblower at the Michigan plant had alerted the Food and Drug Administration to a series of dangerous failures. The factory's equipment had not been properly maintained. The company was releasing batches of formula that had not been adequately tested. Factory managers had covered their tracks by falsifying records.

Running the plant appropriately—replacing aging machinery, conducting additional testing—would have cost money. The imperatives of protecting public health and preserving the company's reputation were locked in a zero-sum contest with the gratification of shareholders. Even as Abbott discovered eight instances of bacterial contamination at its Michigan plant between 2019 and 2021, it dedicated $5 billion to repurchase its own shares.

Any pressure Abbott might have felt to invest in better equipment and operations was relieved by the perks of its market dominance. The company was able to charge American families roughly twice the going rate for baby formula in Europe.

By the time the baby formula shortage seized public attention, the supply chain upheaval had become an American obsession. It was a central part of the explanation for a broader crisis—the stubborn persistence of inflation.

As consumer prices surged by 8 percent compared to the year previous—the highest rate of inflation in four decades—television analysts grimly debated whether this represented a return to the crippling stagflation of the 1970s. Political pundits cast inflation as a dangerous threat to Biden's Democratic Party fortunes ahead of the midterm elections in November 2022.

Implicit in much of the discussion was the assumption that inflation amounted to a head-scratching misfortune that was playing out despite the best intentions of everyone involved. To a considerable degree, this was nonsense. In many cases, inflation represented a triumph of corporate strategy. It was the reward harvested by businesses that had cannily amassed dominance in their markets.

Economists warned that inflation was a pernicious and menacing force that tended to work into the crevices of everyday expectations, prompting labor to demand more pay—a prospect that horrified Wall Street. Once inflation gained momentum, it was very difficult to contain and reverse.

Amid such talk, the Federal Reserve began aggressively lifting interest rates, wielding its traditional tool in the face of rising prices. The logic of this tack had been honed over decades. By increasing borrowing costs for corporations, credit card holders, and home-buyers, the Fed was discouraging spending. That would eventually diminish demand for goods and services, bringing down prices.

Reducing demand would also undermine the bargaining power of American workers while threatening their jobs. But the chattering class and many economists accepted higher unemployment—and perhaps even a recession—as the inevitable by-product of slaying the inflation monster. And never mind that, within lower-income households, joblessness was a far greater peril than higher prices.

Some economists argued that inflation was a temporary phe-nomenon stemming largely from pandemic-related chaos in the supply chain. Once factories got rolling again, they would replace the things that were missing, bringing prices down. Eventually, American basements would be so full of exercise bikes that they could fit no more. That would relieve the pressure on the shipping and trucking industries, allowing things to return to normal. The longer prices stayed high, the less currency these narratives held.

Most of this conversation rested on the acceptance that higher prices were the result of the classic relationship between supply and demand—the basic firmament of economics. The political sphere ar-gued about whether inflation was the outgrowth of reckless spend-ing by the Biden administration, or the consequence of supply chain bottlenecks that were global and therefore beyond the purview of any president. But whatever the explanation, something had thrown supply and demand out of whack. There were not enough goods available to satisfy consumer appetites. That spelled higher prices.

There was some truth to this analysis. Supply and demand nearly always had something to say about prices. But another element tended to get short shrift in the inflation explanation: the conse-quences of extreme market concentration.

The disruptions of the pandemic were playing out in an era in which four conglomerates controlled at least 60 percent of the supply of coffee, bread, cookies, beer, and pork. Within the markets for baby food, pasta, soda, and beef processing, concentration was even greater, with the top four companies owning more than 80 percent of the supply.

Companies with dominant positions in their industries were positioned to exploit the pandemic as an opportunity to lift prices—and far in excess of what they needed to recover their own increased costs for the parts, energy, and labor that went into their production. In short, they fattened their profit margins.

By April 2022, according to one analysis, more than half of the increase in American prices for goods reflected enhanced profits for corporations, while higher pay for workers was responsible for less than 8 percent of rising prices.

Other research revealed that American companies had used the pandemic to mark up their prices vastly beyond their increased costs, to levels that represented the highest on record, with this trend especially discernible in industries in which a few huge companies dominated the market.

When American corporations closed the books on 2021—a year that most families were desperate to forget—they celebrated their sweetest profits since 1950, an increase of 35 percent compared to the previous year. As they detailed their earnings to grateful investors, many executives boasted about their ability to exploit the pandemic as a chance to raise prices while funneling extra cash to shareholders.

In a call with investors in December 2021, Kroger, a major chain of American supermarkets, celebrated higher prices as the key to especially lucrative times.

"We are passing along higher cost to the customer where it makes sense to do so," said the company's chief financial officer, Gary Millerchip.

Here was the explanation for Kroger's whopping 21 percent profit margin during its most recent quarter. The company had also repurchased $1 billion worth of its share over the previous year.

Faced with public anger over inflation and evidence of price gouging, the Biden administration unleashed an attack on market concentration, tasking the Federal Trade Commission with an investigation.

At the center of its offensive was the meat processing industry.

"The meat increases we are seeing are not just the natural consequences of supply and demand in a free market," the White House asserted in a blog post in late 2021. "They are also the result of corporate decisions to take advantage of their market power in an uncompetitive market, to the detriment of consumers, farmers and ranchers and our economy."

This was a legitimate framing of a serious problem, but it was disingenuously packaged as a means of containing inflation. Though antitrust enforcement was woefully overdue in Washington, any effort to redress the abuses of the meatpackers was likely to entail a lengthy legal process. New competitors aiming to construct slaughterhouses to increase capacity—a key administration objective— would require billions of dollars. All of this was likely to take years, providing little comfort to families faced with an immediate spike in their grocery bills.

Still, the mere fact that an American administration was setting its sights on the meat industry represented a potent historical turn.

More than a century earlier, in the era of the Robber Barons, the monopolistic bullying by the meatpackers had stoked public outrage that ultimately prompted the government to intervene and break their chokehold on the market.

The history of the ranchers traced the rise and fall of opportunity and basic fairness in American economic life. For decades, they had benefited from policies that had promoted competition. Now, they were at the mercy of a new crop of monopolists. And their vulnerability along with the concentration of their industry left consumers exposed to shocks to the food supply.

Steve Charter's home—an homage to weathered wood—sat on a bluff affording a commanding view over the plains. The land stretched out unbroken, dotted by sagebrush and buttes, toward horizons more than one hundred miles away.

As Annika prepared a simple yet delicious lunch of grass-fed beef, I noticed a black-and-white painting displayed on the living room wall: a portrait of a cowboy in chaps and a ten-gallon hat, a cigar protruding from his lips, astride a gallant horse.

"That's the outlaw," Ressa said, referring to his great-grandfather.

According to family lore, he had been born in Iowa, then made his way west to Wyoming, where he fell in with the notorious gang run by Butch Cassidy and the Sundance Kid. They robbed banks and railroads, the storehouses of wealth for the detested Robber Barons, and found sanctuary in communities that viewed the bandits as the good guys. Around 1910, the outlaw had opted to go legitimate, using the spoils of his illicit endeavors to buy something real—a cattle ranch in Jackson.

He had entered the industry at a seminal moment in American history. Europeans had pushed Native Americans off their land, decimated the bison herds, and laid down the railroads as they exploited the frontier. Across the West, ranching had burgeoned into one of the most valuable patches of American agriculture. Investment in livestock outweighed the value of all the assets stashed in the country's banks. Cheyenne, Wyoming—ground zero for a speculative frenzy in cattle—was by some measures the most affluent town on earth.

The cattle business was lucrative enough to attract some of the wealthiest families of the era, among them the Rockefellers, the Vanderbilts, and the Whitneys. By 1917, five companies dominated the business of slaughtering cattle and distributing beef. Collectively, they controlled as much as 86 percent of the national market for meat.

Chicago parlayed its status as the nation's rail hub to become the center of this thriving industry. A single stockyard south of the city erected pens large enough to hold 21,000 head of cattle along with 75,000 hogs and 22,000 sheep. So vast was the enterprise that a feeding trough extended for five miles. Here was the place where more than 80 percent of the national meat supply was processed, a stinking zone of dismemberment in which 9 million animals met their messy ends each year.

The Chicago stockyards were the antecedent to Tin Aye's tragic experience, as captured by the muckraking journalist Upton Sinclair, whose 1906 novel, *The Jungle*, revealed the horrors of the slaughter-houses. "All day long the blazing midsummer sun beat down upon that square mile of abominations: upon tens of thousands of cattle crowded into pens whose wooden floors stank and steamed conta-gion; upon bare, blistering, cinder-strewn railroad tracks, and huge blocks of dingy meat factories, whose labyrinthine passages defied a breath of fresh air to penetrate them . . . ," Sinclair wrote. "There were also tons of garbage festering in the sun, and the greasy laun-dry of the workers hung out to dry, and dining rooms littered with food and black with flies, and toilet rooms that were open sewers."

Sinclair was widely credited with bringing the grotesquerie of the Chicago stockyards into public consciousness. This generated pop-ular demands for rules to curb the abuses of the monopolists. But the part of his story that provoked the greatest outrage centered less on the wretchedness of working conditions than the implications for the sanctity of the food supply. Americans were appalled by the revolting places that produced the meat they were serving in their dining rooms.

President Woodrow Wilson ordered a Federal Trade Commission probe of the meatpacking industry. The ensuing report, released in 1919, read like a rap sheet indicting the Robber Barons.

"The producer of live stock is at the mercy of these five companies because they control the market," the report declared. "The menace of this concentrated control of the Nation's food" was "not a casual agreement brought about by indirect and obscure methods, but a definite and positive conspiracy for the purpose of regulating purchases of live stock and controlling the price of meat."

The result of this attention was a landmark law, the Packers and Stockyards Act of 1921, which was aimed directly at preventing the Big Five from dictating prices to ranchers and manipulating the market. The act was part of a broader reorganization in American economic life through which government intervened to root out monopoly power. It was a preview of the New Deal reforms of the following decade, which imposed stringent regulation on finance along with social programs to aid people in need.

This was the backdrop for Steve Charter's father when he arrived in Montana in 1950, having learned that land could be had for as little as $3 an acre. He borrowed money and amassed holdings that at one point reached seventy-three thousand acres.

And this was the world in which Steve and his three siblings learned ranching from their father while attending class at a one-room schoolhouse.

Once a year, the family loaded their cattle into a stock truck, piled into the cab, and drove two hours into Billings, to a sale barn where their animals were auctioned off.

The sale barn was bustling with agents representing slaughterhouses and feedlots from across Montana and the Dakotas, and as far away as Colorado and Nebraska, all looking to buy calves and cows. People jostled around a fenced-in display area as an auctioneer touted the merits of the animals on offer, aiming to provoke a frenzy of interest that drove prices higher. There were so many buyers and so much demand that the Charters could count on fetching a reasonable price.

At a café inside, Charter and his brothers tucked into plates of mashed potatoes and gravy—a treat for kids reared in the isolation of the ranch.

But what Charter remembers most vividly was how his father—a hard-drinking, cantankerous sort—typically drove home content, with a check large enough to cover their bills for another year.

Still, his father counseled Charter to get out, go to college, and forge a different life. He did not see the era of fair competition lasting. He assumed that the packers would resurrect their dominance.

"He saw all this coming," Charter said. "That these family farmers and ranchers wouldn't be viable."

When he was twenty-two, Charter met a woman from Chicago named Jeanne Hjmerstad, who had moved to Montana to work for Northern Plains, the children's advocacy organization. Where he tended to be quiet and cerebral, she was charismatic, pugnacious, and passionate. The romance that blossomed between them was grounded in their surroundings. She was captivated by the ranching life—the horses, the raw beauty of the land.

They were married in 1975. Three years later, his father died, and Charter assumed responsibility for the ranch.

The next year, a former Hollywood actor named Ronald Reagan announced that he was running to become president of the United States.

Reagan presented himself as an American cowboy for his times, donning a ten-gallon hat as he posed for pictures at his own ranch in Southern California. He campaigned as a man of the people who was grounded in the traditional values of a mythologized past ruled by small-town decency. His cowboy shtick drew a stark contrast to the elitist federal bureaucrats he blamed for corralling American liberty. But he governed as a willing tool of corporate interests, entrusting the American marketplace to a new wave of monopolists.

Reagan did not start the wave of deregulation that refashioned the American economy. As we have seen, Jimmy Carter can claim that dubious distinction. But Reagan dramatically advanced the cause. He took a wonky set of arguments grounded in the academic literature and amplified them into a visceral campaign mantra. He demonized government as a bloated, lumbering beast that feasted on squandered tax dollars. He vowed to rescue the country from inflation and economic stagnation by sacrificing federal power to the gods of free enterprise.

"We must put an end to the arrogance of a federal establishment which accepts no blame for our condition, cannot be relied upon to give us a fair estimate of our situation and utterly refuses to live within its means . . . ," Reagan declared in announcing his candidacy in January 1979. "We must force the entire federal bureaucracy to live in the real world of reduced spending, streamlined functions and accountability."

Reagan did not conjure this mission by himself. He installed himself as the mouthpiece for a movement that had gathered strength over decades, quietly yet relentlessly financed by American industry. The intellectual underpinnings for his campaign had been established at academic hothouses like the University of Chicago, whose campus sat less than five miles from the old stockyards. The acolytes of this new faith included the economist Milton Friedman, who dismissed the excesses of the Robber Barons as a "myth" while seeking to eviscerate federal authority.

One key figure at Chicago, the legal scholar Robert Bork, championed the idea that enabled the meatpackers to reconstitute their monopoly power: the lifting of antitrust enforcement.

Traditionally, American antitrust law regarded scale as inherently dangerous. A company with a hold over the market could use it to prey on smaller competitors. It could lower prices to a point where no one could profit, and then lift them once it owned the market. And it could corner the supply of raw materials, making it impossible for new competitors to emerge. This was the cornerstone of the legal foundation that had restricted the scope of corporate mergers, ensuring that no single company could get too big.

Bork asserted that these ideas were not only antiquated but destructive, discouraging innovation. Mergers were to be accepted as the pathway to scale and greater efficiency. The only relevant consideration was what immediately happened to consumer prices. If people could buy goods more cheaply, the government had no business getting in the way of a deal.

This was a radical notion, one seized by American industry seeking license to pursue increasingly consequential mergers.

Reagan was the first president whose government was marinated in this thinking, but he was far from the last. Every administration that came after—Republican and Democrat alike—bought into the concept that industry left to its devices was the best way to produce quality goods at affordable prices. Until Biden's revival of antitrust scrutiny, this was a bipartisan consensus.

The meatpackers gained federal blessing for mergers by promising to revolutionize food production. Greater scale would spread their costs across a larger volume of meat. That would allow them to lower prices.

Wholesale beef prices did edge down slightly until the early 2000s. But in the interim, the largest industry players laid the ground for sharp price increases that unfolded over the following decade. They shut down smaller slaughterhouses, diminishing the capacity to process cattle, and they concentrated their operations at increasingly enormous plants like the JBS facility in Greeley.

Back in 1976, only five American slaughterhouses had been large enough to process at least half a million cattle per year. Collectively, they yielded less than 15 percent of the nation's beef supply. Most of the action was at 145 midsize plants that each had the capacity to slaughter roughly a tenth as many as the giant operations. More plants meant more options for ranchers selling their cattle.

Two decades later, the situation had been refashioned by consolidation. Fourteen plants had the capacity to slaughter at least a million head of cattle per year, and they collectively processed more than two-thirds of the national beef supply.

For the Charters, consolidation spelled fewer buyers showing up at sale barns, and lower prices for their cattle. By the late 1980s, they felt fortunate to break even.

Desperate to inject competition into the business, they joined forces with other local ranchers to revive an abandoned slaughterhouse in Billings. But the packers got wind of their plans. They began paying a premium to area ranchers, locking up the supply of cattle to deny the new slaughterhouse sufficient volume. It soon failed, leaving the Charters nursing a $100,000 loss.

That experience prompted Jeanne to consider another strategy— cutting out the packers and selling directly to consumers. She approached the manager of a grocery store in Billings and gained his permission to set up a stall in the parking lot. She and Steve grilled samples of beef, finding enthusiastic takers.

But the manager soon withdrew his welcome. One of the largest meatpackers had noticed this attempt to bypass its distribution chokehold. The grocery store had to send the Charters away or otherwise forget about gaining deliveries of its beef.

"It kind of took the heart out of me as far as fighting these things," Steve said.

In Washington, the worship of scale remained nearly universal. Through the 1990s, Democrats such as Bill Clinton accepted consolidation within meatpacking—their thinking encouraged by voluminous industry campaign contributions—in the name of making food cheap and abundant to attack poverty.

The following decade brought even greater consolidation, led by the most aggressive player of all: JBS.

The year after it bought Swift, taking control of the slaughterhouse in Greeley, plus eleven other American plants, the Brazilian company handed over $565 million for Smithfield Beef. That deal included a feedlot operation capable of nourishing more than 800,000 cattle at once.

At the same time, JBS announced plans to take over National Beef Packing, the nation's fourth largest beef processor, in a deal valued at $560 million. But that purchase so clearly flouted basic anti-trust law—putting JBS in control of one-third of the beef processing capacity—that the Justice Department filed a lawsuit to block it.

"The combination of JBS and National will likely lead to grocers, food service companies and ultimately American consumers paying higher prices for beef," declared Thomas O. Barnett, who headed the antitrust division at the Justice Department in the George W. Bush administration. "It will also lessen the competition among packers in the purchase of cattle."

Faced with a rare instance of antitrust enforcement, JBS re-lented. The Brazilian conglomerate would have to content itself with controlling a mere one-fourth of the American beef market.

By 2010, the number of beef slaughterhouses had been slashed by 80 percent compared to the late 1970s. A third of the feedlots disappeared, too.

By the time the pandemic arrived, shutting down slaughter-houses, four companies were again in control of 85 percent of beef slaughterhouse capacity, up from 36 percent when Reagan assumed office.

The impacts of the Packers and Stockyards Act had been fully erased.

Not only did the packers reduce capacity to limit demand for cattle, but they reduced what they paid ranchers and feedlots for animals by relying on so-called captive contracts. In these deals, a slaughterhouse committed to purchase cattle long in advance, at prices to be determined later. Unlike at the sale barns, where prices were transparent, such transactions were withheld from public view, diminishing the leverage of ranchers and feedlots.

Increasingly, ranchers like the Charters felt compelled to accept any terms on offer.

By 2019, captive contracts and other secret arrangements were the means by which nearly three-fourths of all cattle changed hands in the United States. Many sale barns had disappeared, and those that remained were sparsely attended.

At Livingston Livestock, a sale barn on the edges of Billings, a dozen buyers looked on from stadium seats one afternoon in early 2022. Mike Hollenbeck occupied a bench in the front row, inspecting the animals being led into the ring as the auctioneer spewed a musical incantation of numbers. He represented several feedlots. Three decades earlier, when he began visiting the sale barn, there were three times as many buyers, he told me.

"The concentration of the industry has pushed people to the sidelines," he said. "It has a depressing effect on the market."

The result was a massive transfer of wealth from the people who raised the animals to the meatpackers and their international investors.

Traditionally, American ranchers took in roughly half of every dollar that consumers spent on beef. By 2020, their share had dropped to an all-time low of 37 cents.

Just as the Federal Trade Commission had found a "definite and positive conspiracy" in the machinations of the meatpackers a century earlier, this sharp reduction in the livelihoods of ranchers was not happenstance. It was the engineered result of decades of consolidation combined with the advent of captive contracts. And the effects were allegedly enhanced by direct manipulation of prices by the packers, according to a class action lawsuit filed by a group of ranching associations.

The lawsuit accused the packers of an organized plot to lower the prices they paid ranchers and feedlots for their animals. The simplest mechanism was limiting their purchases at the sale barns. The prices set there were the key variable in the formulas that determined how much they had to pay for cattle under their captive contracts. By steering clear, they drove prices lower.

But the packers also employed a more aggressive form of price manipulation, according to a confidential witness cited in the lawsuit. The witness—an employee at the JBS slaughterhouse in Cactus, Texas—reported a 2015 conversation with the plant's operations manager in which he learned that the four largest meatpackers had conspired to reduce their purchases of live cattle and slaughter fewer animals. That diminished the supply of beef, which allowed them to charge supermarkets, restaurants, and other buyers higher prices. It also dropped the prices they had to pay ranchers and feedlots under their captive contracts.

I reached out to JBS seeking comment, and they referred me to the North American Meat Institute, the Washington lobby shop. There, I heard from the head of communications, Sarah Little. This was the same spokesperson who had admitted, in internal emails, that the fears of a meat shortage had been stirred up by major processors seeking the Trump administration's help in forcing people like Tin Aye to continue laboring at slaughterhouses.

"Concentration has nothing to do with price," Little told me. "The cattle and beef markets are dynamic."

This was industry propaganda that could not survive a moment's scrutiny beyond the cubicles of Washington lobby shops.

In the hill country of northern Missouri, Coy Young, a fifth-generation rancher, had glumly concluded that beef was a sucker's game.

"The packers control the whole price system," he told me when I visited him in November 2021. "You're feeding America and going broke doing it, and nobody cares. It doesn't pencil out to raise cattle in this country anymore."

Young's modest ranch sat on the edges of the town of Blythedale, population 193. The previous year, he had reacted to weak prices by shifting his focus away from ordinary beef. He had tapped credit cards to borrow $55,000, plowing most of that money into the purchase of advanced artificial insemination technology, aiming to produce a herd of premium breeding cows.

The payoff was supposed to come through a sale that Young hoped would fetch $125,000. But the day of the auction, spooked by the pandemic, traders in the commodity pits in Chicago pushed down the price of live cattle by more than 10 percent. As he trucked his animals to a sale barn, Young felt a sense of foreboding. He drove home that afternoon with an empty trailer, a check for $32,000, and visions of bankruptcy.

A week later, he waited for his wife to leave for her nursing job— their sole means of paying their bills. He found the nine-millimeter pistol that he had given her as an anniversary present and placed it next to his temple.

"You put your heart and soul into something, and then you lose your ass," Young told me. "You don't see any other way out."

Fortunately, his wife had forgotten something. As she unexpectedly pulled back into the driveway, Young put the gun away. He sought professional help for his depression.

But the dismal arithmetic remained. The following year, he sold off his herd and launched a barbecue catering business.

"You're raised a farmer, and that's what you're supposed to do," he said. "It's my family legacy. It's like I'm losing my image as a man."

Steve Charter had largely given up on sale barns by the time the pandemic arrived in early 2020, pushing cattle prices down. He resolved to sell his calves as quickly as possible. He had three rounds to unload, forty head each. For the first batch, he called a man he knew at a feedlot that supplied JBS.

The response left him dumbfounded. The man told Charter that he had to agree to deliver his calves to a JBS plant in Utah on a specific date, with the price to be determined by the packing company on arrival. And he had to commit to those terms immediately or risk no sale.

"If we didn't take that slot, there wasn't any guarantee," Charter told me. "I wanted to tell him to go to hell, but what choice did I have?"

He was hoping to receive at least $1.30 per pound, his break-even point. The word from JBS was devastating.

"Without any consulting or any dealing, they just decided they were going to pay me a dollar a pound," Charter said.

A few weeks later, with the second load ready for sale, he called every potential buyer he knew. No one even wanted to look at his animals, even as many American households were by then panic-buying beef. Slaughterhouses were shutting down, eliminating demand.

He reluctantly went back to JBS. This time, he received only 90 cents a pound—a gut punch.

By the time the third round was ready a few weeks later, Annika found herself contemplating the same path that her mother had forged decades earlier: a bypass around JBS and the rest of the packers.

Jeanne had died in a car accident in April 2011, leaving a void in the Charter home. Annika now tapped into her mother's fighting spirit. She would try her hand at selling directly to consumers, cutting out the middlemen.

Steve was gun-shy, but the status quo promised more losses. He and Annika found a small slaughterhouse that could process their animals. They located cold storage. And then Annika went on Facebook, posting photos of her family, the cows, the plains. "Beef that is all Montana," her post read. "Born. Raised. Cared for and fed in our beautiful state. From our family ranch delivered to your home."

Her post went viral. They sold all forty head, netting an extra $40,000—seeming validation for bypassing the monopolists.

"For the first time, customers' eyes were wide open," Annika told me. "There are bigger powers that are in control of their food supply, and that's a really scary thing. It's all about control. I mean, who do you want controlling your food?"

It was a triumph for the little guy, for the notion of family ranchers, for a meat supply that could operate outside the space dominated by the packers.

Except the Charters still wound up losing money. Fuel was expensive. Grain and fertilizer were soaring in price. And those first sales to the JBS plant had decimated the books.

Which left Steve mulling the heartbreaking possibility that he would have to cash out and try something else.

"There's things that don't have monetary value," he said, tears welling, as the early-afternoon darkness seeped across the land. "So it's kind of the shits that in the end it's just about money."

It was about the money.

Financial interests had shaped the markets that produced nearly everything. The investor class constructed a global production and distribution network that took efficiency as its guiding light. But it was a mutant form of efficiency, one that valued profit margins above all else, even at the expense of reliability. The corporate compulsion for lower costs justified the transfer of manufacturing across oceans at the same time that Just in Time and *lean* demanded a slashing of inventory—a perfect prescription for trouble. The unrestrained celebration of scale placed monopolists in control of once-competitive markets, leaving economies vulnerable to shortages and rising prices once a shock emerged.

And when the pandemic exposed the dangers, emptying store shelves and sending prices soaring while multiplying corporate profits, here was that rare opportunity: a chance to reimagine the whole thing.

PART III

Globalization Comes Home

CHAPTER 18

"WE JUST NEED SOME DIVERSITY."

The Search for Factories Beyond China

Having weathered the shortages of shipping containers, floating traffic jams, the dearth of truck drivers, and other trials of the Great Supply Chain Disruption, Glo's holiday season was, rather miraculously, a success. Elmo and Julia completed their journey across the Pacific, and then to the other end of the United States, validating Walker's agitation and vigilance.

But the experience of moving that single container from the factory in Ningbo to his warehouse in Mississippi left him shaken. His assumptions about globalization no longer seemed secure.

Depending on Chinese factories appeared full of growing hazards.

The tariffs that Trump had placed on Chinese goods were not going away. Biden had effectively extended them indefinitely, while ramping up conflict with Beijing. A new cold war appeared to be unfolding between the United States and China.

Taiwan was a greater flash point than ever, stoking grim speculation that China, under the hardline leadership of Xi Jinping, might unleash its military to seize the island. Biden had added provocation by asserting publicly—and repeatedly—that the United States would come to Taiwan's rescue in the event of attack. That had frayed decades of so-called strategic ambiguity, undermining a policy that had been designed to limit the possibility of confrontation between the two powers. The damage was so evident that the White House had felt the need to publicly clarify that Biden's words signaled no change in the American stance.

Walker was no expert in cross-strait relations, but he grasped that the increased likelihood of a war in the region did not enhance the virtues of relying on China to make and ship his products.

And he was especially concerned about circumstances inside China itself. Xi's extreme Zero COVID policy had not only disrupted factory production but had also sown confusion about national priorities.

For decades, China's leaders had been guided by the imperative to develop the country as quickly as possible, generating tens of millions of jobs for rural people, while increasing the nation's technological prowess and economic vitality. Here was the redress to centuries of vulnerability and despair—first, as a victim of colonialism, and then as a nation besieged by internal ferment. A strong economy would place China in control of its own fate.

That context was the bedrock for international businesses tapping Chinese factories to make goods. There had always been dangers in this undertaking, from the rampant theft of intellectual property to the corrupt intervention of Party officials. But the savings and efficiencies had been substantial enough to justify the risks. So long as foreign investors avoided competing directly with the anointed state-owned behemoths, there was ample space for success. International companies were part of China's economic blueprints, a source of capital and innovation.

But the Zero COVID policy, in which Xi's government quarantined entire cities for months at a time, suggested that fresh priorities were in force. The government evidently was more intent on social control than economic growth. This undermined confidence in China among foreign investors.

Walker and Barker had long contemplated ways to limit their exposure to trouble. Back in 2019, an advisor had urged them to consider the value of spreading their orders across multiple factories in different places, rather than concentrating all their production in a single country.

The pandemic had underscored the urgency of that mission. They were eager to find another place to make Glo's next generation of products.

So, in May 2022, they flew to Ho Chi Minh City, in the south of Vietnam, to scout out new factories. This placed them at the center of a widening current. International companies that had long relied on Chinese factories were increasingly exploring alternatives elsewhere in Asia.

"We just need some diversity in terms of everything that has happened in China," Walker explained.

He liked and respected the people he dealt with in China. He enjoyed his interactions with Zheng and his team.

"We just think our governments are going to fuck everything up," he said. "It's getting really bad."

At first blush, the reexamination of the merits of making products in China appeared to affirm the logic of Trump's trade war. He had explicitly called on American businesses to abandon the country.

"Our great American companies are hereby ordered to immediately start looking for an alternative to China, including bringing our companies HOME and making products in the USA," President Trump had tweeted in 2019. "We don't need China."

Academics tended to indulge a more sanitized term—*decoupling*—to describe the process through which China and the United States were beginning to reduce their mutual economic dependence, as if the two great powers were retreating to separate condos as they filed for divorce.

Few saw decoupling as a comprehensive outcome. China possessed unique scale and breadth of industry that made it next to impossible to replace as the linchpin of global manufacturing. It stood alone in offering raw materials, chemicals, and components, along with factories, port infrastructure, and a population of 1.4 billion people. The Chinese economy was especially dominant in the realm of electric vehicles, a critical element of the global mission to limit the perils of climate change. From mining minerals needed for car batteries to manufacturing such products, Chinese companies were the biggest players. And the United States remained the world's largest consumer market. Whatever came next, China and the United States seemed largely inextricable.

Indeed, by the middle of 2023, American and European officials were indulging a new piece of jargon to describe their designs on an updated relationship with China. They were not interested in decoupling, which they acknowledged would entail expensive economic chaos. Rather, they spoke of a process of "de-risking," in which they would seek to limit excessive reliance on Chinese suppliers, and especially in sensitive areas like medical technology and pharmaceuticals.

Semantics notwithstanding, a marginal separation appeared to be underway.

By the end of 2022, 24 percent of American companies in China were mulling plans to move some of their operations elsewhere, a jump of 10 percent compared to the previous year, according to an annual survey by the American Chamber of Commerce in China. The primary concern among businesses was the degenerating state of relations between the United States and China, with nearly half of all respondents anticipating further deterioration.

A similar share of European companies with a manufacturing presence in China was considering heading for the exits.

Among those who fretted over the minutiae of managing supply chains, the pandemic and its ructions in transportation had compromised the savings of entrusting factory production to faraway places. And they had especially exposed the perils of leaning so heavily on China.

"People were astonished by the complete lack of control over our supply," said Brandon Daniels, chief executive officer of Exiger, a consultancy that managed supply chain risks. "It's an immense political change."

Over the course of 2022, foreign direct investment flowing into China dropped by nearly half compared to the previous year.

Walmart, Amazon, and other major retailers had not suddenly lost their appetite for China's efficiency and low costs. Rather, those virtues now came accompanied by risks that were perpetually shifting and subject to the vagaries of geopolitical relations, making them difficult to anticipate.

There was also the simple fact that China's wages had been rising for years, while its labor force—long treated as infinite—was shrinking. As a result of population control measures, the Chinese workforce was on track to drop by 100 million within two decades.

Then, there were the implications of the evolving tensions between Washington and Beijing. China had secured a dominant hold over key elements such as the processing of rare earth minerals and building blocks for computer chips. That gave its leaders influence over the supplies of crucial materials needed for a host of industrial pursuits, from manufacturing cars and consumer electronics to fashioning fighter jets and advanced weapons systems. The Biden administration had taken this state of play as justification for a series of trade sanctions engineered to slow China's technological development and deny it the capacity to produce its own advanced chips. Any company that depended on China to make technological wares risked placing itself in the middle of the conflict, tripping considerations of American national security.

For companies that sold clothing and other goods that required cotton, relying on China risked accusations that they were profiting on appalling worker exploitation. This was the result of accounts of systematic human rights abuses against the ethnic minority Uyghurs in the Chinese province of Xinjiang—a major source of cotton. The Biden administration accused China's government of genocide in its repression of the Uyghurs, citing reports of forced labor. American sanctions broadly prohibited products linked to Xinjiang from entering the United States. And when multinational companies complied, vowing to avoid Xinjiang cotton, they enraged Chinese consumers, making themselves vulnerable to boycotts.

Reducing the world's reliance on Chinese factories seemed certain to increase prices on a vast range of consumer goods. Multinational companies were in essence trading some of the efficiency that had long defined the era of profit maximization in exchange for diminishing their vulnerability to crises. Instead of purchasing goods in enormous quantities at the lowest prices from Chinese factories, they would spread their business around the globe. They would have to familiarize themselves with the rules, facilities, and power brokers in other countries while staffing new offices. That would add complexity to their operations, absorbing time and money. And that meant that sticker prices at retailers were likely to go up.

Yet even McKinsey was proselytizing about the need for multinational companies to limit their exposure to what it called "structural supply chain fragility," though it pointedly avoided mentioning China. In a survey conducted during the first months of the pandemic, 93 percent of companies engaged in the supply chain told McKinsey that they were planning to boost their "resilience"—the term for lining up alternate suppliers. Some $4.6 trillion worth of trade could be "rebalanced across geographies in the coming years."

The new framing of globalization did not mean that the world was suddenly abandoning China. Indeed, the data revealed that China remained central to the American economy. Despite the bellicose accusations hurled from Washington to Beijing and the tit-for-tat tariffs, the trade of goods between the United States and China exceeded $690 billion over the course of 2022—an all-time record.

But beneath those headline numbers, the Chinese and American economies appeared to be diverging slightly yet meaningfully. Just before Trump had taken office, Chinese-made products had made up

22 percent of total American imports. Five years later, that share had dropped below 17 percent. Among products on which the highest tariffs applied, American imports from China had plunged by more than one-fifth, while those from other parts of the globe had soared by more than one-third.

But this ostensible validation for trade nationalism fell short in one considerable way. Trump had sold his tariffs on Chinese goods as a way to force jobs to return to the United States. Tariffs were "a powerful way to get companies to come back to the USA and to get companies that have left us for other lands to COME BACK HOME," he had declared on Twitter.

Such words were vacuous. The mere act of jacking up prices on imported goods from China damaged the competitiveness of a range of domestic companies, destroying a net three hundred thousand American jobs by the fall of 2019, according to one estimate.

Some companies did take the tariffs along with the disruptions of the pandemic as their cue to remove themselves from China. But this did not presage a rush of production back to the United States. Rather, businesses exiting China tended to seek out other low-wage centers of manufacturing that were insulated from the conflict between the United States and China.

Vietnam emerged as the primary beneficiary of this shift. The Southeast Asian country bordered China, had lower wages, and was following a similar development trajectory, yet had not provoked the wrath of Donald Trump.

By 2022, Vietnam's slice of global furniture exports reached 17 percent, more than double the level of six years earlier, while China's share dipped from 64 percent to 53 percent. Over the same period, Vietnam's share of global footwear exports climbed from 12 percent to 16 percent, while China's dropped from 72 percent to 65 percent.

Apple shifted nearly one-third of the work of manufacturing its AirPods from China to Vietnam. It moved parts of the production of iPads and Apple watches there as well, while transferring some work on its iPhones to India. Samsung shifted most of the manufacturing of its computer monitors from China to Vietnam.

On their face, such moves appeared to limit the risks of depending on Chinese factories. But this frequently proved to be false comfort after a deeper look at the nature of the supply chains. Asian countries that were gaining orders to manufacture products previously made in China were themselves heavily reliant on Chinese factories for parts and materials.

Vietnam's apparel industry imported thread, buttons, and packaging materials from Chinese suppliers while drawing only two-fifths of the needed materials from domestic plants. The fourteen members of a regional trade bloc that included India, Indonesia, Japan, South Korea, Thailand, and Vietnam typically depended on Chinese suppliers for nearly a third of the parts and materials that went into their exports. And that share had increased dramatically since 2010.

The chaos of the pandemic had revealed that one missing component among thousands could be enough to halt production. At its factory on the River Rouge in Michigan, Ford had nearly everything it needed to make its pickup trucks. But without a single key element—computer chips—its finished vehicles were immobilized.

The shifting of production from China to Vietnam and other countries raised the prospect of an unexpected outcome: even greater vulnerability to a shock. Instead of leaning on a single supply chain that was highly concentrated in China, major international brands were scattering their business across multiple countries, raising their costs. Yet they were still faced with the possibility that some future disruption—another pandemic, a natural disaster, or a fresh conflict between Washington and Beijing—could leave them unable to secure what they needed to make their products.

"You still depend on China, it's just that it takes more steps along the way," Brad Setser, a former US Treasury official and now a fellow at the Council of Foreign Relations, told me. "There's more places where things could go wrong."

Jacob Rothman, who ran the factory that made Glo's products in Ningbo, had spent most of his adult life in China. He had successfully treated the country like a one-stop shop for every conceivable part and material. He and his wife, a Chinese national, were raising their son in Shanghai. But the trade war between the United States and China, followed by the chaos of the pandemic, had altered Rothman's grasp of the future. Every potential customer was suddenly intent on avoiding the pitfalls of entrusting its supplies to China.

"Everyone demands that you provide other options," Rothman told me.

He had opened a factory in Cambodia. And he had established joint-venture plants with local partners in Vietnam.

These were the factories that Walker and Barker were going to visit as they sought to limit their exposure to China.

The story of investment leaving China for Vietnam was far from new, though it had in recent times accelerated dramatically.

In 2005, I had visited several factories in Vietnam that had been established by Chinese entrepreneurs seeking relief from wages that were rising at home. For the Chinese overseers, Vietnam beckoned as a country much like their own in the early years of its market reforms.

"We have no problem understanding that bribes have to be paid to get things done," said Zou Qinghai, a Chinese textile entrepreneur in Hanoi. "Vietnam's way of developing is simply a copy of China's."

Vietnam's government had since invested aggressively in highways, electrical grids, ports, and other basic infrastructure, outfitting the country to seize the opportunities of the export trade.

Vietnam joined the World Trade Organization, and then signed a trade deal with the European Union. Between 2014 and 2022, Vietnam's growth in exports shipped over long distances increased nearly fivefold.

Long before Trump affixed tariffs on Chinese goods, forward-looking American companies were beginning to shift production from factories in China to Vietnam. Nike was an early adopter. Columbia Sportswear and Adidas set up plants. Then the electronics trade began to move.

And then came the trade war, followed by the pandemic, upending Chinese industry and ports.

This was how Walker and Barker found themselves in a vibrant city named for another Communist revolutionary, Ho Chi Minh, in a country focused on supporting multinational businesses in the enterprise of making money.

Walker and Barker stayed at a glittering hotel, its lobby boasting a Murano glass chandelier, its marble bathrooms bedecked in orchids. They explored alleyways lined with boutiques. And over the course of a mere two and a half days on the ground, they visited seven factories, guided by a local representative for Platform88.

The factories appeared modern and capable, run by ambitious managers who spoke excellent English and seemed eager for their business.

"I was really impressed," Barker said. "It was a really high standard."

They especially appreciated that the factories were all under Platform88's umbrella. That meant they could deal with the same people they already knew and trusted, relying on them to find the needed parts and materials, rather than having to start over in a country they had visited for the first time.

But when they dug a little deeper, the merits of shifting to Vietnam grew complicated.

Barker and Walker had brought with them prototypes of Glo's latest generation of merchandise—colorful jars that lit up when filled with water. These products needed no electronics, only plastic, which meant that they could be made easily in Vietnam. But the cubes—the source of most of Glo's revenues—did require electronics, and that posed a challenge.

They were sitting around a conference table with one factory owner who assured them that he could make the cubes. But then he let slip that he would have to import the electronics from China.

"Well, that doesn't do us any good," Walker replied. "When we want to produce something in Vietnam, we'd like to only produce it when we're not sourcing parts from China. Because that's the whole point of diversification."

But that was an exceedingly difficult demand. For two decades, Chinese manufacturers had been shifting their own capacity to Vietnam. Many of the local factories were managed by Chinese companies and drew on parts exported from China.

Still, Walker and Barker saw a partial solution to their problem. They resolved to continue making their cubes in China, where the supply chain was vastly more developed, while entrusting plants in Vietnam with their newer, simpler items. Here was a first step in a gradual process of broadening their horizons.

They flew home intent on starting production in Vietnam by the summer of 2023 at the latest, giving them just enough time to make and ship their Christmas stock for the year.

Then, nothing happened. When Walker asked Rothman and Zheng to quote prices for orders at the Vietnam factories, he was told that the numbers would have to wait until the plants could configure their machinery to yield samples. Weeks stretched into months without clarity.

By early March 2023, Walker was both confused and worried about the calendar. He spoke to Zheng.

"I'm like, 'We need to be moving on this,'" he said. "'We're getting down to crunch time.'"

Zheng explained that Platform88 was concentrating its investment in Cambodia, where it wholly owned a factory. Its plants in Vietnam were merely joint ventures with local companies, meaning that it was not fully in control of operations and had to share the profits with its partners. Cambodia also had lower duties on importing parts from China. And with pandemic restrictions now lifted, Rothman was able to visit the Cambodia plant and accelerate plans to upgrade and expand it.

For all these reasons, Platform88 was concentrating its focus and investment on Cambodia, while deemphasizing Vietnam.

Walker and Barker began to wonder if they were wasting their time on something that was not going to happen anytime soon. What was the point of having flown all the way to Vietnam if those factories were not real options? They were game to explore moving their orders to Cambodia, but Zheng counseled against that, too.

"He was very blunt," Walker told me. "He said Cambodia is about ten years behind where China is in terms of understanding injection molds."

Zheng's recommendation was to simply keep making everything in China. Later on, they could revisit shifting the work to Cambodia.

Walker saw no other option. The whole exercise of seeking to diversify Glo's production had brought home the simple fact that real alternatives were few and often illusory. Manufacturing continued to orbit China.

"It's terrifying," Walker said. "You look at half the things on my desk and most of them are going to have a sticker on the bottom saying 'Made in China.' China has become so good at sourcing."

He had recently needed to track down a new version of an LCD screen for a display case for Glo's products on retail shelves. He went on Alibaba, the enormous Chinese e-commerce site, and submitted a request for the product. Within minutes, he had twenty replies with detailed quotes, pictures of factories, and online reviews posted by previous customers. He selected one, paid by wire transfer, and was expecting his order to be completed within three weeks.

"In Vietnam, that doesn't exist," Walker said.

That didn't exist almost anywhere. In the rest of the world, what he had pulled off with a single query on Alibaba would have required weeks of old-school networking, looking for factories, calling, texting, and emailing to solicit quotes, and then waiting.

Businesspeople often talk of barriers to entry—the difficulties of new competitors entering a given marketplace, from the costs of building a factory to the challenge of securing raw materials. Companies that have built successful brands see barriers to entry as a useful form of protection.

Barker and Walker had discovered that Chinese manufacturing effectively had barriers to exit. You could try to make your stuff somewhere else, but good luck. The same forces that had turned the country into the center of global manufacturing tended to reinforce that status.

It's very hard to break up with Chinese industry, I suggested.

"Think about how crazy that is," Barker said. "Because we're not even breaking up with China. We're just asking if we can also see other people at the same time."

CHAPTER 19

"GLOBALIZATION IS ALMOST DEAD."

Bringing Factory Jobs Home

In other parts of the manufacturing realm, some American companies were taking hostilities between Washington and Beijing as the impetus for a whole new kind of production—one centered on making goods domestically.

Taylor Shupe was not the most obvious candidate for inclusion in the nascent movement to return factory production to American shores. Reshoring, as the concept was known, was all about bringing jobs home. He had dedicated most of his adult life to sending them away.

Raised in Southern California, Shupe had begun studying Mandarin when he was only fifteen, seeking to prepare himself for his plan to one day run a global company. During a college semester in China, he had focused on scoping out factories that could make products for the business he launched even before he graduated—a company that sold protective cases for laptop computers.

Then he oversaw production for a start-up company called Stance, which designed and manufactured premium socks adorned with bold colors, surfer patterns, and price tags reaching $25 a pair. The socks gained the adulation of prominent tastemakers—Rihanna, Will Smith, LeBron James.

Like most apparel brands, Stance relied on a Chinese factory to produce its wares. Even after he left the company in 2017 to start his own sock company, FutureStitch, Shupe held on to the Chinese operation, constructing a new factory to make his products.

Then, labor costs rose in China. Trump slapped tariffs on Chinese imports, and Biden continued them. In Washington, two major political parties that agreed on almost nothing achieved consensus that China represented a mortal threat to the American way of life.

By the time the pandemic arrived, multiplying the costs of transporting goods across the Pacific, Shupe was already feeling urgency to make his socks closer to his customers.

In the summer of 2022, he opened a new factory in Oceanside, California, about forty miles up the coast from San Diego. When I visited the plant early the following year, only twenty people worked there, using machinery to apply decorative designs to blank socks imported from China. But Shupe planned to more than double the workforce by the end of the year while shifting production to California.

"We're headed to a state of hyperlocalization," he told me as we zipped down the freeway in his Tesla at alarming speed. "The big disruptions that have occurred over the past three years have definitely exposed the sort of risk that we didn't think existed. Which brands want to set up new supply chains in China now? The political risks, they're the thing that everyone is talking about now."

As a relatively small producer of niche goods, Shupe's company did not have a sprawling supply chain to manage. And still the implications of the changes underway in the world economy were forcing him to grapple with enormous variables. The era of globalization that had shaped the first phase of his entrepreneurial life had centered on

China. The next phase, already unfolding, was influenced by evolving hostilities between Washington and Beijing. This posed complex logistical, financial, and legal challenges.

Shupe had to factor in the pitfalls of continuing to rely on socks from China, given the prevalence of cotton harvested in Xinjiang and the threat of sanctions against companies that traded in that commodity. More broadly, he recognized that China had become a branding liability.

He was first and foremost an entrepreneur, a creature of the fashion world who sought to cater to rapidly changing consumer tastes. He and his fellow start-up founders had correctly divined that high-end socks were a retail frontier waiting to be exploited; a long-neglected mass commodity that could be elevated into a platform for individual expression. But tastes were influenced by factors that included political sentiments. When you asked the customer to buy your brand as a statement of identity, you invited scrutiny of the values governing the enterprise.

Shupe took no satisfaction in seeing the United States and China devolve into a state of hostility. He admired the resourcefulness and drive of the people he had worked with in China, and he appreciated the sense of limitless possibility. But he also understood that the Americans whose feet he aimed to win over were increasingly prone to viewing Chinese industry as unsavory, and even malevolent.

He had achieved success in part through a keen understanding of the potency of social media and celebrity in driving consumer impulses. The most effective marketing entailed the product gaining a cameo in an authentic story. No traditional advertising could rival the power of an Instagram post revealing an NBA legend donning Stance socks, or the Jay-Z song that celebrated the glories of the brand. ("This ain't gray sweat suits and white tube socks / This is black leather pants and a pair of Stance.")

What Shupe grasped keenly was that Chinese production had become a damning detail in the story of his product, whereas manufacturing things in the United States had been transformed into the makings of a new narrative, one in which the customer was on the right side of history. They were not only investing in American communities but also responding to climate change by limiting the carbon emissions entailed in shipping containers across the ocean.

This was not something that could be achieved by shifting orders to Vietnam or some other country in Asia. Those options still entailed the orders crossing the Pacific to reach American consumers.

The shipping industry was a leading source of pollution blamed for climate change, emitting 3 percent of global greenhouse gas emissions.

Under rules imposed by the International Maritime Organization—the United Nations arm that regulated shipping—the industry was required to upgrade its fleet to limit those emissions. Complying with the new rules would demand investments reaching $1.5 trillion over the next three decades. That was likely to increase the cost of shipping, further challenging the equation of making goods in Asia for sale in North America. It was certain to intensify scrutiny of the environmental impacts of container shipping, posing reputational risks for international brands.

"Consumers want to know where things are made more than ever," Shupe said. "And how things are made."

He had staffed his California factory to yield a satisfying answer to that second question, too. FutureStitch had partnered with the local government to hire formerly incarcerated women, most of them Black and Latino.

People like Tasha Almanza, a mother of four who had served time for selling drugs, were at the center of the brand's story.

"We are women working together," she told me. "We're here to empower each other. This has given me the opportunity to rebuild my life."

Henry Ford had constructed a business that cast his employees as customers. Their pride in workmanship combined with their satisfaction behind the wheel was a source of word-of-mouth marketing long before social media. Shupe was forging a tale in which he invited his customers to participate in advancing American progress by employing people eager for a second chance.

"When you think about the employees that we're hiring, who just so happen to be the most forsaken of any other employment group in the United States, and their stories of struggle, this is real power," he said. "And everything else coming from China, not only is it hollow of those sorts of social elements, it's a net negative politically."

You could buy into that framing, or you could react to it skeptically, divining an opportunistic marketing pitch. Either way, its existence signified a shift in the American conversation.

Most of the attention to reshoring was centered on weightier concerns than socks or light-up Elmo dolls. Huge, capital-intensive industries that involved considerations of national security or future technological prowess—like advanced chipmaking, electric cars, and pharmaceuticals—dominated the action.

Trump had fulminated about the need for American companies to abandon China and return home. But most of the production leaving China landed in other low-wage countries in Asia—not only Vietnam, but also Malaysia, India, and Bangladesh.

Still, Trump did succeed in boosting American manufacturing in one key area. He convened a federal initiative known as Operation Warp Speed that reserved supplies of key ingredients and equipment needed by pharmaceutical companies as they successfully developed lifesaving COVID-19 vaccines in record time.

The Trump administration gave $1 billion to Johnson & Johnson and $1.5 billion to Moderna to support their expansion of factories in the United States. It handed out contracts to key suppliers like Corning, the glassmaker, to ensure that enough vials were on hand to permit large-scale production. Trump invoked the wartime Defense Production Act in claiming authority to direct supplies as needed, giving pharmaceutical companies assurance that they would have the ingredients required to make their vaccines at American plants.

Once in office, Biden expanded on those efforts, accelerating the availability of COVID-19 vaccines by coordinating domestic suppliers. And he dramatically intensified the American campaign to constrain China's economic development. He left in place Trump's tariffs on Chinese imports, and he opened a new front in the trade war: computer chips.

In August 2022, Biden signed into law a bill marketed by its acronym—the CHIPS Act, which stood for Creating Helpful Incentives to Produce Semiconductors and Science. The centerpiece was $52 billion worth of direct subsidies and tax credits lavished on companies that agreed to produce chips at plants in the United States.

Administration officials touted the law as the key to weaning American industry off its troubling dependence on chips forged in Asia while promoting national self-sufficiency. Here was liberation from the vulnerability of relying on factories in Taiwan and the perpetual risk of a military invasion from China.

The money was aimed at bolstering domestic technological capacity as a way to prevent Chinese industry from gaining an advantage. Companies that accepted the subsidies as they constructed chip plants in the United States were explicitly barred from expanding their factories in China for at least a decade.

"It's no wonder the Chinese Communist Party actively lobbied US business against this bill," the president declared at the bill signing ceremony.

Strikingly, the bill captured the support of seventeen Republicans in the Senate. This rare display of bipartisanship attested to the extent to which both parties had come to view China as a profound threat.

But it also indicated something else: the political power of catering to the shareholder class.

The $52 billion the government showered on the semiconductor industry came accompanied by flexible limits on what the recipients could do with the money. They were free to carry on buying back their shares and extending dividends, so long as they did not use grant funds directly for that purpose—a restriction that could be bypassed by any creative accountant. Cash was being distributed by the taxpayer for the enrichment of shareholders in the name of buttressing national security.

Two months after signing the CHIPS Act, the Biden administration followed up with fresh policy directives engineered to deny China the ability to advance its own chip industry, especially those with military and surveillance applications. A series of orders from the Commerce Department severely restricted American companies from exporting advanced chips known as graphic processing units, and limited sales of semiconductors that could aid China's supercomputing enterprises. American companies were explicitly required to seek licenses before they could supply Chinese firms with machinery that could be used to make advanced chips.

The Biden administration was not content to simply stop American technology from reaching China. The new directives included rules aimed at preventing companies anywhere on earth from supplying Chinese companies with chips that could be harnessed for supercomputing or artificial intelligence purposes, provided they were made with American technology.

Through diplomatic channels, the White House pressed key allies—especially the Netherlands and Japan—to codify their own restrictions barring companies from selling Chinese firms the machinery used to make advanced chips.

By then, the thrust of the policy was evident. Biden was intent on limiting China's technological prowess, and especially in areas that verged into military capability, employing a containment strategy that had no pathway to de-escalation. What had begun as a trade war had widened into a conflict that some were describing as a new cold war.

That same year, Congress approved a sprawling bill that unleashed $370 billion worth of spending and tax credits aimed at spurring the transition to cleaner forms of energy to address climate change. The package included up to $7,500 in federal tax credits toward the purchase of electric vehicles. The measure came with stipulations designed to encourage American production of electric cars.

One requirement spelled out that the tax credits applied only to cars whose batteries were made with large volumes of minerals harvested in the United States or one of its allies—at least 40 percent of the value initially, with that share doubling to 80 percent by 2027.

Later that year, the Biden administration gave a direct push to the domestic production of electric vehicles, announcing $2.8 billion in grants to finance battery factories along with projects to extract minerals.

A century earlier, as the combustion engine emerged, Henry Ford had sought to amass a dedicated supply of raw materials to ensure his ability to make growing numbers of cars. As a new supply chain took shape to support the next generation of cars—those powered by electricity—the Biden administration was adopting a similar role on a national scale.

Some economists decried the Biden administration's embrace of Made in America provisions. They would increase prices and squander capital that could be more efficiently deployed to buy goods from foreign suppliers. They perpetuated the fantasy of national self-sufficiency, which was attainable nowhere. Every region of the globe relied on trade with another for a minimum of one-fourth of its supply of at least one critical good.

And the concerted campaign to boost American manufacturing furthered the deterioration of the rules-based international trading system, solidifying the reality of the new era in which economic might and nationalist brio defined the course of global commerce. The campaign upset key alliances, especially in Europe, where companies saw in Biden's rules a protectionist attempt to subsidize American ventures to the detriment of fair play and European competitiveness. Yet European companies were themselves constantly on the prowl for new subsidies and trade protections, and the Biden administration handed them a giant opportunity: if the Americans were doing it, then the Europeans had to do likewise or surrender sales.

Some warned that a narrow obsession with boosting American industry would come at the expense of critical efforts to address climate change. China was the dominant source of low-cost solar panels. If the United States pursued policies designed to boost reliance on domestic solar while limiting Chinese imports, that would delay the process of transitioning to cleaner energy by increasing the costs. Extra factory jobs in Ohio would be paid for by exacerbating the ravages of droughts, storms, and rising seas around the globe.

There was validity to all these concerns, and ample cause to worry about the consequences of sacrificing international trading rules to the project of reviving American industrial fortunes. Yet decades of faith in the rules-based trading system and China-centric globalization had produced a situation that was simply untenable. The world's wealthiest country had been reduced to begging for computer chips while scrambling to make medical devices in the midst of a pandemic. The norms of global trade had created the conditions that had advanced climate change.

Something had to give.

Some economists parsed data to argue that international supply chains, contrary to being the cause of pandemic-related shortages, were in fact the solution. Trade across oceans had allowed the United States and the rest of the globe to replenish depleted stocks of medical gear. But that reasoning was tautological. Yes, once you dispatched production far away, slashed inventories, and then suffered shortages, it made sense to use whatever capacity existed to make more of what was missing.

Whatever professional trade experts had to say, the active encouragement of reshoring had become a full-blown political event that was altering the business climate. The combined effect of the Biden administration's policies was an industrial construction boom across the United States.

Chipmakers led the way, expending eye-popping sums to establish and upgrade American factories. Intel, the Silicon Valley giant, announced plans to invest $20 billion to erect two chip plants in Ohio, creating seven thousand construction jobs. Micron, another American company, outlined plans to spend at least $100 billion over the following two decades to construct a complex of chip factories near Syracuse, New York.

Taiwan Semiconductor Manufacturing Company, the world's largest manufacturer of advanced chips, detailed plans to invest $40 billion to expand a factory outside Phoenix. The cost of building the plant was likely to run four to five times that of a similar facility in Taiwan. And the chips produced in Arizona were likely to cost half again as much as those made in Taiwan. This was the price of adjusting to the geopolitical realities altering international trade, shifting the supply of chips needed by American industry away from an island that could, at any moment, come under assault from China.

At a ceremony to commemorate the installation of the tools for the plant, Morris Chang, the ninety-one-year-old founder of TSMC, cited the Arizona investment as an indication of the world economy being reshaped by the imperatives of reshoring.

"Globalization is almost dead," Chang declared.

By the end of 2022, the semiconductor industry had dedicated almost $200 billion to construct and expand forty factories engaged in chipmaking in sixteen states, while generating forty thousand future jobs.

A similar wave of construction unfolded in the electric vehicle industry. In 2022 alone, companies announced $73 billion of investments in factories making electric car batteries in the United States. By the end of the year, Ford was breaking ground on a new $5.8 billion battery plant in Kentucky.

As enormous new factories took shape across the United States, the buildout catalyzed demand for a broader American supply chain that could deliver the basic elements of industry, from power transmission gear to construction equipment.

"We have a once in a century opportunity," said Aamir Paul,

president of North American operations at Schneider Electric, which manufactured fuse boxes and other piece parts for the management of power. "Semiconductor investments are going to create a ripple effect coupled with reshoring and supply chain resilience."

In other words, once you started making computer chips in the United States, that would demand new power plants, requiring fuses and wiring and computer terminals. And once you built up that capacity, that would alter the economics of all manufacturing, making the United States a more attractive proposition. Eventually, entrepreneurs like Hagan Walker and Taylor Shupe would find themselves able to purchase raw materials, components, and technological know-how at home.

There were fundamental reasons to doubt how rapidly this could play out, or how far it would go. The United States was woefully behind most European countries in terms of training workers for cutting-edge manufacturing. It lacked the needed electricians, plumbers, and other skilled trades. It would take years to amass the labor force along with the raw materials and machinery required to build out even the projects that had been announced.

"We absolutely do not have the capacity for all of that now," Paul told me.

Indeed, by late 2023, Taiwan Semiconductor was delaying plans for production at its new Arizona factory while complaining that it could not hire enough skilled workers. And reshoring was largely driven by federal subsidies, not by the market. It was the result of a concerted campaign by the Trump and Biden administrations to alter the map governing international trade, as opposed to the product of dispassionate economic calculations.

"The national security concerns, and the geopolitics with Taiwan, those factors are motivators," said Eskander Yavar, a managing partner at BDO, an international business consulting firm. "If there's no subsidies in place, I think reshoring becomes a slower roll."

Yet this was the very point of the subsidies, Yavar added. They changed the economic equation for the executives presiding over the supply chain, making the United States a more appealing place to invest. And they created pressures for everyone to go along. Once a rival business availed itself of subsidies to construct an American chip plant, gaining proximity to domestic customers and insulating itself against geopolitical crises, any company that stuck with sole reliance on Asia was effectively at a competitive disadvantage.

But outside of obvious beneficiaries like chipmakers, few industries saw the appeal of setting up in the United States. In a suddenly hyped conversation around the revival of American industry, talk of reshoring was widespread, yet actual instances were few. Bringing factory work home to American shores was challenged by the same consideration that had sent it away in the first place: lower prices could be found elsewhere.

An advocacy group called the Reshoring Initiative launched by Harry Moser, a retired manufacturing executive, had long labored in obscurity. When I caught up with him in February 2023, Moser was testifying on Capitol Hill, pressing the imperative to quit depending on China by bringing production home. He cited data showing that one-fifth of all products imported from China could be made in the United States without raising costs, after factoring in the savings from avoiding shipping.

His reports catalogued a tsunami of investment, with 364,000 new American jobs in the pipeline thanks to reshoring via foreign investment in 2022 alone—an increase of more than half compared to the previous year.

But the claims seemed to exceed the reality.

One of the companies included in Moser's database as an example of successful reshoring was a Texas-based start-up called Volcon that made electric bicycles, motorcycles, and off-road vehicles. Volcon was not making any of its products in the United States, its CEO, Jordan Davis, told me, though he was scoping out prospects for American suppliers of parts.

All of Volcon's e-bikes were made in China. The company was intent on keeping the retail prices of its bikes below $3,000. Making them in the United States would have cost roughly twice that.

"You can't ask someone to spend six grand on a bicycle," Davis said. "The infrastructure is not here for the kind of high-volume, low-cost manufacturing that exists in Asia."

The question was whether and how quickly the subsidized investments flowing into electric vehicles, pharmaceuticals, and computer chips would crystallize a broader revival of the American industrial base.

But one factor was clearly trending in that direction—the political and sentimental value of manufacturing goods in the United States.

Shupe had grown up in San Juan Capistrano, a classic, sun-splashed chunk of the Orange County coast, tucked between the megalopolis of Los Angeles to the north and laid-back San Diego to the south.

Like most of his friends, he had surfed and skateboarded as a kid, giving him intimate familiarity with the sartorial concerns of people who strove to be cool and comfortable at the same time.

From the time he was eight, Shupe was deep into one entrepreneurial venture or another.

"Every Christmas, I'd ask for things that would allow me to generate money," he recalled.

One year, it was a lemon crusher that allowed him to sell lemonade to the neighbors. Other years, he received a cotton candy machine, a sno-cone maker, and a rock polisher—all harnessed to yield marketable products.

At twelve, Shupe was selling boxes of chocolates and trinkets door to door. Then he took a job as a delivery boy at a florist run by a man from Taiwan, using the experience as an opportunity to learn rudimentary Mandarin, positioning himself to get to Asia.

The fifth of seven children, Shupe was reared as a devout Mormon, a faith he later rejected. Growing up, he took it as a given that he would deploy somewhere as a missionary. His stint with the florist allowed him to include on his application the fact that he spoke some Mandarin. In 2002, when he was eighteen, the church dispatched him to southern Taiwan.

Shupe has since come to view the mission as a colonial enterprise draped in cultural supremacy. But he seized it as an opportunity to achieve Mandarin fluency. The experience engaged his business instincts. The church grasped the motivational powers of gamification, employing metrics of assessment—how many Books of Mormon placed, how many people enticed to the faith.

"The objective was to convert," Shupe said. "I became very competitive."

Two years later, he returned to the United States and enrolled at Brigham Young University. By the time he landed in China for an exchange semester at Nanjing University, he was eager to line up a supplier of neoprene for his business making protective sleeves for laptops.

He found a factory in southern China, in the industrial city of Dongguan. Back in Utah, he wired $10,000 to the supplier for his first order—enough for five thousand laptop cases.

In the winter of 2006, his shipment arrived at the Port of Long Beach in a twenty-foot container.

Shupe drove to a shipping yard with a U-Haul truck to pick up his products and was horrified to discover that cool weather had caused the neoprene to densify and shrink. None of the cases fit. He shaped a wood block into a replica of a laptop, stretched the cases over it, and then wielded a hair drier to melt and mold the material into the appropriate form.

The business grew. Circuit City, the national electronics retailer, became his largest customer. But when the chain disappeared into bankruptcy in 2009, Shupe was left with a massive stash of unsold product along with $250,000 in debt. He liquidated his inventory and shut down the business.

The same year, he joined with three other entrepreneurs to start Stance. They initially focused on skateboarders, using stretchy material that applied light compression to stop them from sliding down calves, and employed designs from local artists.

They contracted with a factory outside Shanghai to make their products. Shupe supervised production, initially flying back and forth every few weeks between California and China. But the lack of daily supervision yielded trouble. Machinery mysteriously disappeared from the factory. Orders got screwed up amid long-distance communication problems.

After six perpetually jet-lagged months, Shupe moved to China and set up home near the plant.

When he started FutureStitch, he held on to Stance's China operation, constructing a new factory on the same site.

From the beginning of his new brand, Shupe had intended to set up a factory in the United States to get production closer to customers. The trade war and the pandemic accelerated the timetable.

FutureStich began with contracts to make socks for Stance and other brands. Every month, it shipped roughly two dozen forty-foot containers from the port of Ningbo to Southern California. Not only were the costs of transportation multiplying, but the time needed to get products to market increased from three to ten weeks.

This was especially troubling given Shupe's fixation with customized goods. He was pursuing plans to quickly release socks with photographic images of key highlights in sporting events—the game-winning shot in the NBA Finals, the triumphant horse at the Kentucky Derby. He aimed to supply customers the power to use the Web to special order socks for Valentine's Day, for birthdays.

None of that was enhanced by waiting for weeks for containers to cross the Pacific.

"You look at the moment, the heat of the meme," he said. "By the end of the month, it's not even a tenth what it was."

The breakdown in global shipping exposed the pitfalls of Just in Time, generating pressure for companies to boost inventory to prevent running short of product. But as the Federal Reserve began lifting interest rates in early 2022, that increased the cost of borrowing to expand inventory. All of which argued for making products closer to customers.

Here was the logic behind the factory in Oceanside.

Shupe's interest in social justice combined with more prosaic staffing considerations prompted him to forge a partnership with San Diego–area government agencies to recruit women who had spent time behind bars.

Many employers avoided hiring people with criminal records, viewing them as risky. Shupe saw in these women a powerful drive to succeed. They had demonstrated extraordinary resilience in surviving their circumstances. They were subject to regular, government-administered drug tests. And they were uniquely invested in their jobs.

"We have to be here, or we go back to jail," said Almanza, one of the first hires at the factory. "It's our freedom that's on the line. We're working for a bigger purpose, because we're trying to change our lives."

A single mother of four in her midforties, she had previously been a phlebotomist, drawing blood in a San Diego–area hospital, where she earned nearly $28 an hour. Then a man she worked with began stalking her, she told me, cornering her when they were alone in a lab. When she filed a sexual harassment complaint, she was fired for poor attendance, and never mind that she had missed shifts out of fear for her safety.

Without a job, Almanza lost her apartment. She and her children moved out just after Christmas 2020, first cramming into a mobile home parked in a friend's driveway, and then bouncing between cheap motels.

She had long ago struggled with drugs but had been clean for seventeen years. Sinking into depression, she relapsed, and then resorted to dealing meth in order to keep her family fed, she said.

She was arrested and charged with conspiracy to distribute in June 2021. She spent two months in a county jail, and two more at a federal detention facility. She pleaded guilty in exchange for a sentence of a year of supervision.

At the government office where she applied for a cash assistance program, someone told her about FutureStitch. She applied and was hired at $20 an hour. She moved her family into an apartment and was soon promoted to supervisor.

Almanza piloted a forklift across the factory floor, moving boxes of blank socks to the printing area. She participated in yoga classes taught by the head of human resources.

"My inner self is healing," Almanza told me—not the sort of words typically spoken about a factory job.

Thanks to consultancies like McKinsey, *flexibility* had become a euphemism in American commercial life, a license for bosses to schedule shifts with little notice. FutureStitch turned the equation around. Employees were free to notify the company only a day in advance on their availability for work, allowing them to juggle family responsibilities. They were excused to attend meetings with probation officers.

FutureStitch was building a skateboard park on the grounds to allow mothers with older kids to bring them to work. The company was offering mindfulness meditation classes.

Shupe acknowledged that part of the appeal for hiring formerly incarcerated women was the impact on branding—"a marketing tool," he called it. But he also saw benefits to being surrounded by single mothers, who had insights on everyday problems, like the casual torture of putting shoes and socks on children with other ideas. This was the genesis for Shupe's latest obsession, a product that was a cross between a shoe and a sock—a piece of footwear that had a strong sole, yet could be worn by itself, and tossed in the washing machine.

On the morning of my visit, Shupe convened his design team in a small conference room to examine a prototype for the new product.

Shupe was excited by the fact that it could be produced on a circular knitting machine. That meant it could be constructed fast and would have no seams to rub against. It was to be made of cashmere for comfort and warmth, along with a breathable, stretchy material. It had no laces. Its outsole was to be made of Vibram, which was made in the United States from recycled materials. The whole thing could be built through no more than five stages of manufacturing, as compared to the eighty or ninety entailed in some footwear.

The simplicity made it a perfect candidate for reshoring.

"It has all the right formulas for 'Made in the USA,'" Shupe said. "This is about creating something here in the United States with interesting design. We'd have to story-tell about this."

Yet there were limits to how far this could go in the United States, Shupe acknowledged. He saw reshoring not as a comprehensive future but as a marginal shift in the making of physical products—"a ten percent portfolio allocation," as he put it.

There were still too many materials that were unavailable in the United States, too many items that were much cheaper in Asia.

A private venture like his could take the long view and spend extra to expand in the United States. But publicly traded companies were perpetually stuck contending with the pressures of the next quarter. They were captive to the permanent imperative to seek lower costs.

For many companies, a middle path was required. They needed a way to avoid exposure to the uncertainties of the ocean, and especially to China, by making goods on the same shore as the customer. Yet they still had to limit their costs.

This was the situation not just for American businesses, but for manufacturers around the globe.

The shock of the pandemic was prompting companies to think regionally. Chinese manufacturers would increasingly depend on satellite operations in Southeast Asia. Western European companies would expand their already considerable presence in Eastern Europe and Turkey. American companies would set their sights closer to home.

This spelled a great opportunity for one country in particular: Mexico.

CHAPTER 20

"OKAY, MEXICO, SAVE ME."

How the Global Supply Chain Turned Its Back on the Water

Isaac Presburger was astonished by the size of the order.

His apparel factory outside Mexico City had been making uniforms for Walmart for several years, but never more than a few thousand at a time.

Suddenly, in February 2022, Walmart was asking for fifty thousand uniforms in one shot—an order worth roughly $1 million.

Presburger oversaw sales at Preslow, an apparel business started by his great-grandfather, an Ashkenazi Jew who emigrated to Mexico from Eastern Europe. He had carefully cultivated Walmart, his biggest customer. He had recently made pilgrimage to the company's headquarters in Bentonville, Arkansas, in pursuit of additional business.

Still, the fifty-thousand-unit order wildly exceeded his expectations. Preslow's rising fortunes underscored how the contours of the global economy were being reconfigured, creating a potentially enormous opportunity—not just for his family clothing business, but for Mexico and the rest of Latin America.

Walmart, the world's largest retailer, had long drawn heavily on factories in China to make its goods. The pandemic had exposed the risks of depending on a single country separated by an ocean from its most important market, the United States.

Unlike Walker and other niche players, Walmart could afford to charter its own container ships, but those vessels were still getting stuck in the floating traffic jams off major American ports. China's zealous Zero COVID policy was disrupting factory production and shutting down shipping terminals. And there was nothing to be done about the American tariffs on goods imported from China, no antidote to the unmistakable reality that Beijing and Washington were locked in a fierce struggle for supremacy.

Walmart was so enormous, its revenues exceeding half a trillion dollars over the course of 2022, that it could conduct itself nearly like an imperial power. It did not need to dispatch buyers around the globe to scout out suitable factories. It simply waited for factories to send their supplicants its way, granting audiences to supplier representatives who descended on the northwestern corner of Arkansas to pitch their offerings.

The people going in to see Walmart these days were coming away with one emphatic message. Getting products on the shelves of Walmart's stores required that they have a comforting answer to the question of where the goods were being made.

Two decades earlier, a company that manufactured its wares somewhere other than China had little to no chance of selling to Walmart. Here was a clear indication that the supplier could not possibly be offering the lowest price.

Now, exclusive dependence on Chinese factories had the opposite effect, signaling vulnerabilities in the supply chain, and bringing meetings to an abrupt halt. Just as Walker had gone to scout factories in Vietnam, Walmart was intent on drawing goods from a diverse array of countries. And the company was especially keen to bring production closer to customers, an approach known as nearshoring—a middle-ground alternative to reshoring. Walmart was not trying to move factory production back to the United States en masse. That would have dramatically increased the costs of its wares. Rather, it was seeking to limit the distance between the factories making its goods and its American customers, while still relying on relatively low-wage countries.

Mexico fit the bill.

Even before the upheaval of recent times, moving a shipping container full of goods from a factory in China to a warehouse in the United States took a minimum of six weeks. Transporting the same load from anywhere in Mexico to the hardest-to-reach destinations north of the border required at most two weeks.

This was why Presburger's factory north of Mexico's capital was suddenly humming with activity.

Seamstresses leaned over their sewing machines stitching up uniforms, underneath paper cutout streamers draped from the rafters, as loudspeakers blasted out *ranchera* music.

"Walmart had a big problem with their supply," Presburger told me. "They said, 'Okay, Mexico, save me.'"

On the other side of the border, in a quiet bedroom community north of Dallas, Jose and Veronica Justiniano were grappling with their own troubles.

Their family embroidery company made uniforms for surrounding businesses, including their most important customer, Gloria's Latin Cuisine, a chain of fine-dining restaurants in the major cities of Texas that was famed for its delectable black bean dip.

The Justinianos bought plain white uniforms from a huge distributor that imported them from Asia—mostly China and Vietnam—and then embroidered the logos using machines installed in an extra bedroom of their home.

Amid the disruptions of the pandemic, their supplier was chronically short of many items, including chef's coats and linen blouses donned by waiters. They were disappointing their own customers by delivering late—a threat to the upward trajectory of their lives.

Born and raised in El Salvador, the Justinianos had left behind a terrifying civil war in their home country to achieve middle-class security in Texas.

Jose had landed first in Los Angeles in 1992 when he was only nineteen. He joined his father, who had been living off itinerant construction jobs, standing on street corners at dawn in the hopes of securing work.

Jose had slept on a cot stuffed into a closet at his father's apartment while working as a janitor at the Beverly Hills jail, and then as a billboard installer, climbing narrow staircases to paste up advertisements above the roar of the freeways.

A friend in Dallas told him there were better jobs there, so he moved, taking a position as a dishwasher, and then as a cleaner at a hospital. He worked nights while studying English at a community college in the daytime.

He got an associate degree in science, parlaying that into an entry-level job at an auto parts factory where he stayed on for eighteen years, rising to supervisor. He met Veronica, who was taking care of an elderly couple. She had worked with her mother in a sewing shop back in El Salvador, and she had the idea for their business.

In 2018, they bought their first embroidery machine, using it to fashion logos for local unions, a landscape gardening company, a janitorial operation, and restaurants.

The chaos of the pandemic jeopardized their growth. Their supplier of uniforms had typically delivered their orders within a day. By the middle of 2020, with container ships from Asia stuck off ports, they were waiting for months.

Jose began searching for other suppliers. He stumbled on a website for a family-owned company based in the Mexican city of Guadalajara—Lazzar Uniforms.

Lazzar's commercial director, Ramon Becerra, then in his late thirties, was digitally savvy and skilled in the art of search engine optimization. He had engineered his site to capture attention from

Spanish-speaking people in the United States, cognizant that this was a group exceeding 40 million.

"We know the US is the future for us," Becerra told me when I visited him in Guadalajara, a vibrant city famed for its Spanish colonial architecture. "We're integrating with the region. It's something that is happening."

Lazzar's team of designers occupied a light, airy room, gazing at computers as they conceived customized versions of uniforms. Outside, on a minimalistic production floor, workers fed spools of fabric into cutting machines that sliced the material into strips. The company delivered these swaths of fabric to sewing centers scattered throughout the country, where seamstresses fashioned them into shirts, pants, and jackets.

Becerra's team conferred on the particulars of what the Justinianos desired: a light fabric that vented away moisture, providing relief from the heat of the kitchen. The two companies were able to communicate easily in Spanish by phone and video, without having to deal with a time difference.

The Justinianos ordered a few dozen chef's jackets. By the fall of 2021—just as Hagan Walker's container was marooned off Long Beach—Veronica's Embroidery was purchasing one thousand linen shirts in a single order, at prices close to what its previous distributor charged for imports from Asia.

Late the following year, the Justinianos flew to Guadalajara to visit Becerra. They discussed a potential partnership in which Lazzar would set up a warehouse in Texas, with Jose handling American distribution.

"This year has been a wakeup call for the US," Jose told me. "We have to reconsider where we get our stuff made."

Becerra was excited by this prospect, but also cautious about the pace of growth. Seventy percent of the fabric he used to make his clothing was imported—from Thailand, Colombia, India, and China. It would take years and vast sums of money for Mexican textile operations to take shape.

He frequently turned down orders, intent on not committing to more business than he could handle.

Lazzar was talking to Ford about supplying uniforms for its workers at a plant in Tennessee. It was in discussions to outfit a construction company in Florida and a landscaping business in California. But Becerra had passed on the prospect of making prison jumpsuits for a distributor that supplied them to institutions across the United States. He had offered to make them for $27 each. The distributor countered at only one-third that amount.

"I literally told them, 'Go to Bangladesh or Ethiopia,'" Becerra said. "We want to be a socially responsible company."

He was not interested in the traditional race to the bottom, seeking to beat the China price. Rather, he was focused on boosting his quality and design skills while exploiting the benefits of accessibility to the largest economy on earth.

He was confident that Mexico and the United States were embarked on a grand-scale joint venture, a collective effort to adapt to a new form of globalization in which China was no longer the sole hub of manufacturing.

"Mexico has to rise up to the opportunity," Becerra said.

The notion that Mexico had become the American fix for the pitfalls of globalization was rich in irony.

Three decades earlier, the Texas magnate Ross Perot had mounted a US presidential campaign largely defined by fearmongering about the "giant sucking sound going south" as Mexico hoovered up American factory jobs.

Unions in the United States had long decried the North American Free Trade Agreement, or NAFTA—the trade bloc comprised of Canada, Mexico, and the United States—as an affront to American job security and wages.

More recently, Trump had disparaged Mexico in crude and racist terms, promising to erect a wall along the border to keep out immigrants he derided as "rapists" and drug dealers. He called NAFTA "the worst trade deal ever made" while renegotiating its terms.

But the breakdown in transportation across the Pacific combined with the trade war between China and the United States had altered Mexico's place in the American conversation. Swapping goods across the Mexican border made solid economic sense. It generated demand for parts and products fashioned in the United States by American workers, and far more so than trade with China.

"The reality is that Mexico *is* the solution to some of our challenges," Shannon K. O'Neil, a Latin America specialist at the Council of Foreign Relations in New York, told me. "Trade that is closer by from Canada or Mexico is much more likely to create and protect US jobs."

By the time the pandemic upended transportation crossing the Pacific, the supply chains of Mexico and the United States were already intertwined and growing more so. Some 40 percent of the value of Mexico's exports to the United States consisted of parts and raw materials that were made at American plants. By contrast, only 4 percent of the value of imports from China was American-made—a number that Beijing appeared intent on driving down through government-directed industrial policy.

In San Diego, an entrepreneur named Raine Mahdi had launched a start-up called Zipfox that served as matchmaker between American companies in need of factories and Mexican plants seeking new customers.

Mahdi had previously run a cardboard packaging business. He had depended on factories in China, gaining an intimate understanding of the risks of sending payment across oceans and then waiting for delivery.

"Everybody who sources from China understands that there's no way to get around that Pacific Ocean, there's no technology for that," he told me. "There's always this push from customers: 'Can you get it here faster?'"

Over the course of 2022, Mexico and the United States exchanged goods worth nearly $780 billion, an increase of 45 percent compared to 2020, the year the pandemic began. Over the same period, the trade of goods between the United States and China grew roughly half as fast, reaching $690 billion.

It was not that Mexico was replacing China. Rather, an incremental yet meaningful shift was underway. As companies scrutinized their supply chains for vulnerabilities, they were opting to balance out factories in Asia with new plants that provided closer access to the American market.

During the first wave of the pandemic in 2020, Mexico's proximity literally proved a lifesaver. The city of Tijuana, just over the border from San Diego, was already sending more medical equipment to the United States then China. As imports from China succumbed to delays, factories across Mexico stepped up to supply American hospitals with ventilators, high-grade masks, and other critical gear for frontline medical workers.

In the Texas border town of Laredo, the busiest overland port in the United States, the products clustered in warehouses attested to how manufacturers on both sides of the border were joined in a collective enterprise.

Inside one facility that managed southbound cargo, boxes of auto parts sat on pallets, freshly arrived from factories in Nebraska and Iowa. They awaited trucks that would carry them two hours south, to distribution centers in the Mexican city of Monterrey, and then on to auto factories throughout the country. Components for heavy equipment made at a plant in Illinois were destined for Irapuato, in the center of Mexico.

Black fifty-five-gallon drums full of universal primer concocted in Ohio were en route to a paint manufacturer in Monterrey. Orange plastic crates held seals for car doors made at a factory in Alabama, on their way to an assembly plant in the Mexican state of Jalisco.

Nearby, at a companion warehouse dedicated to freight moving north, brake parts for truck trailers manufactured in Monterrey awaited air shipment to trucking companies in Texas, South Dakota, and Canada.

Laredo was sometimes dismissed as an overgrown truck stop. A tangle of roads led to a pair of bridges spanning the Rio Grande, the meandering border separating Texas from Mexico. Belching tractor trailers passed through at all hours, dropping off and picking up cargo at the warehouses carved into the scrub-covered desert. From the dilapidated hotels to the hitchhikers gathered on the edges of every gas station, Laredo felt like a place that most people were just passing through.

Yet swelling volumes of goods were also passing through— roughly $800 million worth every day by late 2022. That number seemed certain to increase as multinational companies shifted pro- duction from China to Mexico. Laredo was poised to become an in- creasingly vital center of international trade.

On a crisp, sunny morning in December 2022, Laredo's elected leaders convened at City Hall to celebrate the achievement of a mile- stone. Federal data had just been released showing that $27 billion worth of goods had moved through the local port in October, exceed- ing the flow through the twin ports of Los Angeles and Long Beach.

The then mayor, Pete Saenz, spoke a mixture of Spanish and English as he stood in front of a banner displaying the city seal—the flags of Mexico and the United States side by side, along with the words *Gateway to Mexico*.

Laredo was perfectly positioned to handle the logistics of near- shoring, he told me. This trend was only gathering force.

"A substantial amount is the fear of ending up with a stoppage or impact to the supply chain," the mayor said. "We can't be totally dependent on Asia, and China in particular."

But the exuberance over Laredo's opportunity came tinged with anxiety about whether local infrastructure was about to be over- whelmed by the growing numbers of trucks crossing the Rio Grande.

"We've got to get ahead of this tsunami that's coming," the mayor said. "We're behind now."

A massive buildout was underway. North of the city, an army of excavators tore at the pale soil, turning ranchland dotted by cactus into industrial parks, warehouses, and trucking yards on both sides of Interstate 35, the ribbon of pavement linking the border to the midsection of the United States and Canada.

Kansas City Southern, the American railroad, had recently broken ground on a $100 million project that would double the capacity of a rail bridge spanning the Rio Grande.

Two million square feet of warehouse space was under construction in Laredo and its broader environs, an increase of 5 percent. Just as Southern California's Inland Empire had sprouted with warehouses that were effectively an extension of the docks in Long Beach, these new structures were part of the infrastructure for a shifting of the supply chain to Mexico.

Yet the warehouses in Laredo were already at 98 percent occupancy, much like the facilities in Southern California, raising questions about whether the town could handle what was coming. Trucks were backed up, often for hours, waiting their chance to clear customs inspections on the World Trade bridge, the primary span extending over the Rio Grande.

A local developer was pursuing plans for a new bridge to be constructed just south of Laredo, a $360 million project. He had gained the permission of the Mexican government while expressing confidence that he would soon secure clearance from the US State Department.

This was, at best, an uncertain proposition. In Washington, accommodating future flows of trade took a back seat to the treatment of the Mexican border as a wild and menacing frontier. Though the Biden administration celebrated nearshoring as a partial solution to excessive dependence on Chinese factories, that putative concern had yet to extend down to Laredo.

At the City Hall event, the mayor appeared with officials from the Mexican state of Nuevo León, on the opposite bank of the river. They detailed plans to aggressively upgrade highways reaching the border, while making greater use of the Colombia bridge, a lightly trafficked span crossing the river twenty-five miles northwest of Laredo. They spoke of a singular devotion to security in borderlands besieged by violence linked to the movement of drugs.

Nuevo León was led by a brash, thirty-five-year-old governor, Samuel García, who was betting his legacy on the courtship of foreign investment to boost jobs. The state was home to the largest Lego factory on earth. Tesla, the electric car maker, was finalizing plans to construct a new factory.

The capital of Nuevo León, the city of Monterrey, was a boomtown full of upscale restaurants, luxury goods outlets, and newly erected high-rise office towers in various stages of completion. The city was home to Tec de Monterrey, an elite university often referred to as Mexico's MIT, the linchpin of visions to boost innovation toward the creation of a local supply chain that could support advanced manufacturing.

As the governor described it, the state's development plans were being accelerated by the refashioning of the global economy, and especially the evolving hostilities between the United States and China.

"Nuevo León is having a geopolitical planetary alignment," he told me as we sat across a wooden table in his office at the government palace—a warren of grand rooms with high ceilings, terracotta floors, and balconies looking out on a cobblestone plaza and beyond, to the jagged peaks of the Sierra Madre.

In the fifteen months since García had assumed office, nearly $7 billion in foreign investment had poured into Nuevo León, making the state the second largest recipient after Mexico City. The governor was days away from departing for the World Economic Forum in Davos, Switzerland, where he would woo more foreign money.

His pitch centered on the utility of the North American trade deal. A company could set up a factory in Nuevo León, assemble goods, and then ship them by truck or rail to the United States and Canada for sale free of duty.

García had a powerful ally in his campaign to attract investment.

Major manufacturers with factories in the United States were demanding that their suppliers erect nearby plants.

I ran into a visiting delegation from a South Korean company that produced parts used in construction machinery. Among its largest customers was John Deere, the American maker of tractors and construction equipment. Deere had a factory in southern Texas. It was no longer willing to entrust its inventory of parts to the vagaries of ocean shipping. Its suppliers either had to set up plants that were linked to Texas by road or rail, or risk watching a competitor capture Deere's business.

The Korean company's North American sales manager was in Monterrey scoping out possible factory sites. He saw in Deere's stance clear evidence of a redrawn global economy. By then, shipping prices had plunged to levels close to where they had been before the pandemic. The ocean carriers were again worried about excess capacity and willing to move containers across the Pacific for less than $1,500. Still, he assumed that the push for suppliers to establish factories closer to their customers would prove permanent.

The reconfiguration underway was about more than shipping prices. It was about the pitfalls of concentrating production in Asia. It was a recognition that the United States and China appeared irretrievably bound for conflict, just as Russia's assault on Ukraine brought home the possibility that the world was splintering into rival camps. International businesses were recalculating the impacts of a volatile mix of factors—natural disasters, climate change, future pandemics. Academics were debating whether a process of deglobalization was underway.

"After going through the pandemic and the supply chain crisis, the China COVID shutdown, many North American manufacturers would like to eliminate the risk as much as possible," the Korean sales manager told me. "Globalization has ended. It's localization now. Globalization only works when geopolitical situations are stable. There certainly has been a break point where globalization is a risk."

———

Globalization was not really ending. The merits of international trade remained unimpeachable for the simple reason that no part of the planet was anywhere close to self-sufficiency. Nonetheless, a fundamental shift was palpable. The type of globalization that had prevailed for decades was clearly up for renegotiation. China was no longer the default, one-stop center for the world's manufacturing needs. And the very concept of a global supply chain was giving way to a series of regional hubs.

In the new configuration taking shape, the nations of Southeast and South Asia were likely to serve as suppliers to manufacturers in China that would produce goods primarily for the Chinese market. Companies in Western Europe would ostensibly rely on factories in Eastern Europe, Turkey, and Africa. Consumers in the United States would depend on manufacturers clustered closer to home, especially in Latin America.

This was a needed and sensible adjustment to the excesses of Just in Time and the lopsided reliance on supply chains that stretched across oceans. The world would be spared some of the emissions spewed by container ships. Companies like Walker's would eventually gain options for manufacturing that were much closer to home, reducing their exposure to future crises. Consumers would face less risk of disruption.

Columbia Sportswear had been early to move production from China to Vietnam, setting up an office there in 2001. It had accelerated that shift after Trump launched his trade war. But as countless other brands opted to do the same to bypass the American tariffs on Chinese imports, Vietnam's ports and industrial zones were growing congested, the price of labor and land rising. So, in October 2023, a half dozen Columbia executives took a reconnaissance trip to seek out factories closer to the company's most lucrative market, the United States.

Columbia was focused on Central America, seeking to take advantage of a regional trade deal. Under its terms, clothing made at factories in the region could be sold in the United States free of duty provided that the yarn used to make the fabric was produced at an American mill or within Central America.

Stan Burton, Columbia's global head of apparel manufacturing, had spent three decades in the business overseeing factories in Asia for Nike, Under Armour, and other prominent brands. At six foot five, with a raffish grin, Burton was a hulking and jovial presence. He was now tasked with limiting the risks of heavy reliance on faraway plants. Only 7 percent of Columbia's global production was then centered in Central America. Burton aimed to double that share over the next three to five years.

"We're really repositioning from Asia," he told me as we toured a factory in Guatemala City. The previous day, he had taken a look at a pair of plants in El Salvador.

Making clothes in Central America generally cost 5 to 10 percent more than in Vietnam, Burton estimated, but that was before factoring in the expense of shipping, to say nothing of the time required for transit. Moving a container of goods to the Port of Seattle from Vietnam generally took a month. The same cargo could be shipped from Guatemala in only a week. And that shorter duration would allow Columbia to replenish its stocks far more quickly, which meant that the company could safely go lean on inventory, seizing the benefits of Just in Time.

After the factory tour, Burton and his fellow Columbia Sportswear executives drove into the mountains to the old colonial capital of Antigua. They had lunch at a hotel whose dining room glistened with polished hardwood. They explored cobblestone lanes, sampled the local coffee, and posed for pictures inside the ruins of a half-built cathedral abandoned to earthquakes. When they walked through a leafy plaza anchored by a fountain, some of the team stopped to examine handicrafts laid out by Indigenous women—colorful bracelets, earrings, woven ponchos. Burton tried to shoo them away. "You know those are all made in China," he said.

The clearest evidence that the geography of manufacturing was changing lay in the fact that Chinese companies were themselves constructing factories in Mexico, seeking to protect their unimpeded access to the lucrative American market.

Chinese companies were making use of the newly reworked North American trade deal. So long as they satisfied rules requiring that they use a minimum percentage of parts and raw materials produced somewhere in the region—the precise share varied by product—they could label their wares "Made in Mexico" and ship them duty-free into the United States and Canada.

A Chinese company called Lizhong, which made aluminum wheels for the auto industry, had just broken ground on a huge factory outside Monterrey, its first plant in North America. Its presence in Mexico was the result of pressure from its customers. Lizhong sold wheels to Ford and General Motors. They were insistent that their supplier set up a plant near its own factories in the United States.

Hisense, a Chinese maker of home appliances, had recently opened a pair of gleaming factories. I was intrigued to encounter the company in Mexico because, back in 2004, I had visited its first plant in Europe, a modest factory erected in an unremarkable town in western Hungary.

Hisense had begun operations during the late 1960s, in the midst of China's disastrous Cultural Revolution, churning out cheap AM radios to broadcast Chairman Mao Zedong's mission of class struggle. It had since advanced into one of the country's largest manufacturers of televisions while setting its sights global.

To establish its Hungary plant, Hisense had assigned an executive who spoke no Hungarian and only minimal English. He had arrived in the town of Ostffyasszonyfa with a suitcase full of instant noodles, rudimentary designs for a new factory, and a canny understanding of the changing shape of international trade. Hungary had recently joined the European Union, gaining inclusion into its common market. That made it an ideal jumping-off point for the rest of the continent. Hisense could ship in components from China, assemble televisions in Hungary, and then sell them free of duty from Greece to Germany to Ireland.

Two decades later, the company was essentially employing the same model in Mexico. Its factories were making refrigerators and stoves for the American market.

The industrial park where the Hisense factories sat was the brainchild of a Monterrey corporate lawyer named César Santos. A decade earlier, he had been approached by a developer in Los Angeles who was representing a Chinese electronics company then looking to set up a factory in Mexico. Santos controlled a key asset—a 2,100-acre parcel of land only 150 miles from the border.

The parched brown terrain had been his family's cattle ranch when he was a child, the scene of family horseback-riding adventures. Now, he saw an opportunity to cash in on the changing configurations of the global economy.

Santos flew to China to meet with his potential Chinese partners. He rode a high-speed train from Shanghai to the lakefront city of Hangzhou, the headquarters for the Holley Group, which had already constructed an industrial park for Chinese companies in Thailand.

Santos, the Holley Group, and another Chinese partner, the Futong Group, forged a joint venture company in 2015. So began the Hofusan Industrial Park. They planned a grid of warehouses and factories fronted by a hotel and temporary apartments for visiting managers, plus more than twelve thousand homes for workers along with schools, a hospital, and shops.

The Holley Group dispatched Jiang Xin to oversee the new project. He had previously worked at the company's operation in Thailand. Mexico presented an altogether different proposition.

"Chinese companies had no idea about Mexico, and the only things we knew were bad things, dangerous things," Jiang told me.

"Then," he added, "Trump came."

The trade war eroded the economic benefits of making goods in China for sale in the United States. Suddenly, Jiang's phone was vibrating with texts from Chinese companies that were contemplating Mexico as they sought easier access to the American market.

"Chinese companies wanted more options," Jiang said. "And we are one of their options."

A Chinese company that made artificial grass set up a factory. So did a Chinese maker of pumps used by public fountains. Kuka Home, a Chinese manufacturer of sofas, built a plant. By the middle of 2021, all but one of the twenty-eight factory sites slated for development in the first phase of the park had been claimed.

Then, another Chinese furniture maker, Man Wah, inquired about the final parcel.

Man Wah operated scores of factories in China, from the frenetic industrial zones in the southern province of Guangdong to the northern city of Tianjin, using them to produce inexpensive recliners that it shipped around the world.

Trump's tariffs had prompted the company to buy a factory in Thailand while constructing another in Vietnam, using those plants to serve the American market. But the soaring price of shipping had beggared that strategy. Man Wah was moving 3,500 forty-foot containers per month from Vietnam to the United States. The costs had increased as much as tenfold.

Bill Chan, a Man Wah executive based in Shenzhen, used the Chinese social media platform, WeChat, to reach out to Jiang. His questions were direct. How large was the available lot? (Eighty-four acres.) How were the local highways? (Not great, but improving.) Were there any authentic Chinese restaurants in the area? (None whatsoever.)

Within weeks, Man Wah had committed to purchase the lot, and Chan became the chief executive officer of the company's Mexico subsidiary. He had never set foot in the country, let alone the industrial park, where he swiftly resolved to invest $300 million to construct a factory.

"Our main market is the United States," Chan said with a shrug. "We don't want to lose that market."

In January 2022, he met with a representative of the Holley Group

at Shanghai's Pudong airport. There, he signed the contract to purchase the industrial park site before boarding a flight to San Francisco,
en route to Monterrey, leaving behind his wife and two children.

Man Wah was so intent on establishing production in Mexico
that it leased a small factory to make sofas while the larger plant
was still being constructed. Even before Chan nailed down the temporary site, he had Man Wah fill seventy shipping containers full
of machinery and raw materials in China and send them across the
Pacific to Mexico.

"We always do things quick," Chan said. "Don't worry about anything. Just do it."

There were a few significant details worth worrying about, though.

Man Wah needed to hire six thousand workers to staff its new factory. It had been accustomed to managing people in places like China
and Vietnam, where labor unions were essentially banned, and factories could count on what felt like a bottomless supply of laborers.
Mexico was a democracy with a thriving labor movement. The unemployment rate in Nuevo León was about 3.6 percent. Man Wah and
the other Chinese factories were competing with North American,
European, and South American companies for a limited set of hands.

Retaining workers would depend on accommodating their needs,
perhaps supplying minibuses that could bring people from the villages where they lived. The company would need to throw barbecues
at holiday time, and hand out bonuses when employees' children
were born. These were considerations that managers like Chan
grasped. Yet he had to answer to bosses in China, who tended to see
such line items as gratuitous—a tension unlikely to fade.

The biggest question—not just for Chinese companies, but all
manufacturers in Mexico—centered on how quickly a domestic network of suppliers could emerge to support increased production.

In Monterrey, I visited a cutting-edge factory constructed in
recent years by Lenovo, the Chinese company that had purchased
IBM's computer business two decades earlier. Lenovo was using the
plant to make servers, the racks of electronic boxes that store data

used by financial companies, technology firms, and others engaged in so-called cloud computing. Workers slotted electronics into boards, completing single servers in as little as forty minutes. All of the production was sold in the United States and Canada, replacing goods that had previously been made in China.

The new factory had opened in 2020. For the first two years, the company had continued to rely on Chinese factories for one key electrical component—so-called motherboards. But it had recently located a supplier in Guadalajara and shifted its orders there. It had ceased importing cardboard packaging from China, entrusting this business to Mexican suppliers.

Still, for many key items—from memory circuits to specialized cables—Lenovo was likely to continue drawing on suppliers in Asia for years, and perhaps decades.

"There's no supply chain for these things in Mexico," said Lenovo's Western operations director, Leandro Sardela, a Brazilian based in Monterrey.

Yet even on these items, the chaos of the pandemic years had prompted a foundational change. Lenovo had redirected its historic faith in Just in Time in favor of building up inventory.

The company had long been ruled by a kanban-like system inspired by Toyota. It kept on hand only enough parts and raw materials to satisfy its confirmed orders. But the delays in cargo across the Pacific had exposed the risks of going too lean. Lenovo had supplemented three warehouses in Monterrey with two more to ensure it would not suffer future delays because of parts shortages.

"Our competitors are doing the same," Sardela told me.

The world had run out of enormous quantities of goods, from the frivolous to the lifesaving. The result was a fundamental reordering of global business, a redress to the cultlike reverence for lean production and faith in supply chains stretching around the planet. China's central dominance in manufacturing was giving way to a focus on regional supply networks.

Just in Time was—at least for the moment—being supplemented

by Just in Case. But there was no guarantee that this would stick. Just in Time had combined with global supply chains not because of social concern or considerations for the customer, but because this was an efficient way to cut costs and reward shareholders. The Great Supply Chain Disruption had produced huge changes, but the interests of shareholders remained paramount. After a few years, when the pandemic finally receded into memory, lean inventories would likely return.

But the emergence of regional hubs—as opposed to making everything in China—showed promise as a way to limit the risks. From its inception, Toyota had insisted that its suppliers were clustered close to its factories to ensure that it could quickly replenish parts. If American factories became dependent on suppliers that were closer at hand, the risks of minimal inventory would be tamed.

Nearshoring held the potential to stabilize Just in Time.

The breakdown in the supply chain was more than a configuration problem. The value of human labor had been downgraded over decades. The people delivering packages, maintaining railroad systems, and driving trucks were supplying American households with goods in exchange for wages that left their own families bereft of basic economic security. Though pay had risen in many industries amid acute labor shortages during the pandemic, many workers remained in close proximity to calamity—one unexpected illness or car breakdown removed from a devastating reckoning. And the basic insecurity of the people doing the work left the supply chain itself insecure, prone to breaking down in the face of a shock.

As Henry Ford had presciently warned a century earlier, a low-wage business was a business perpetually vulnerable to suffering shortages of workers. If that had not been sufficiently clear before, the pandemic made this truth impossible to ignore.

Major retailers and transportation companies justified their commonplace forms of exploitation by taking credit for satisfying the demands of the most sacred group of all: consumers. This framing rendered the interests of consumers and shareholders as one and the same. Both benefited from keeping wages low and workers insecure. Both were threatened when workers demanded better wages. Higher pay meant greater costs, which diminished profits and increased the price of goods.

No one was more obsessed with serving consumers than Ford. He presented the automobile as an instrument that would unleash new forms of consumer power, a portal to fresh experiences that would prompt people to open their wallets. But he also grasped that workers and consumers were the same people, their interests indivisible. Most of the consumers arriving at Ford dealerships to drive home new Model Ts depended upon paychecks to finance their spending. Wages had to be bountiful or the whole system broke down.

By contrast, tens of millions of American workers today earn so little that they cannot afford childcare or medical expenses, to say nothing of a new car. And their corporate employers justify their meager pay as the means of keeping prices low for their customers— a wholly separate group of people.

So long as normalcy continues to depend on the desperation of working people, the supply chain is perpetually at risk of descending into chaos. The ultimate security depends on stable pay and working conditions for the people tasked with keeping the gears turning.

You can shift factory production from China to Vietnam or Mexico without altering the relationship between the shareholder class and those who live on paychecks. You can satisfy the standards of consultancies like McKinsey in their newfound appreciation for resilience purely by drawing on a greater variety of suppliers. But restoring resilience, yielding a supply chain that is truly reliable, requires a whole other level of change.

It demands a redrawing of the bargain between labor and shareholders.

Ordinary workers must regain a critical element that has long been in chronically short supply: the means of earning a decent living.

CHAPTER 21

"PEOPLE DON'T WANT TO DO THOSE JOBS."

Robots and the Future of Shareholder Gratification

On an overcast morning in September 2022, I walked into a cavernous convention hall in Philadelphia to survey the future of the supply chain as conjured by those running the businesses that deliver the goods.

I was attending a trade show called Home Delivery World, a conference and display space for hundreds of companies demonstrating increasingly sophisticated means of moving products from warehouses to front doors.

The executives assembled inside the Pennsylvania Convention Center had been chastened by the upheaval of recent years. They had left themselves vulnerable to no end of trouble—from shortages of truck drivers and warehouse workers to excessive reliance on Just in Time. They offered assurance that they were not flinching from the tough but unavoidable lessons.

What was needed going forward was a "people first approach," declared Amiee Bayer-Thomas, the chief supply chain officer at Ulta

Beauty, a major retailer of makeup, skin care, and hair products, as she delivered a keynote address. "People first in everything."

Yet out on the showroom floor, human workers were generally depicted as problems to be minimized and costs to be contained.

People came accompanied with considerations that did not serve the bottom line: needy children, spouses demanding time, illnesses, hobbies, and vices. People posed challenges to the optimal scheduling of work—the flexibility celebrated by McKinsey. And people were a source of resistance to a future centered on automation.

Much of the display space was dedicated to technological offerings that diminished the need for people and their profit-diluting foibles. Robot manufacturers demonstrated their latest models, touting them as efficiency-boosting augmentation to warehouse workers. Staffing companies beckoned with services promising to make it easier to eliminate full-time employees, replacing them with part-time gig workers. Driverless trucks and drones would assume central responsibility for completing the final leg of the supply chain journey.

"It's hard to get people motivated to do this work," said Kary Zate, senior director of marketing communications at Locus Robotics, a manufacturer of so-called autonomous mobile robots—basically, carts that roll through warehouses, accompanying humans who pluck goods off shelves. "People don't want to do those jobs."

Here was an appealing depiction for the executives wandering the showroom floor, one concocted to absolve them of responsibility for the breakdown of the supply chain. The fact that warehouses were understaffed while containers sat on docks for a lack of drivers was not their fault. It did not reflect their reluctance to increase wages or improve working conditions. Rather, employees were to blame. They had apparently lost their will to work.

At the dawning of mass assembly and the emergence of the American consumer market, Henry Ford had responded to the fear that he might run out of motivated workers by doubling their wages. At Toyota, Taiichi Ohno had warned that the pace of assembly had to be regulated to ensure that the slowest workers could keep up.

Yet the following century, as those responsible for the supply chain confronted dysfunction in the economy, they were pursuing the opposite approach. They were leaning on flexible staffing as the way to maintain production schedules on their own terms, and without having to increase pay. They were embracing robots as alternatives to calibrating their operations to the mortal limits of human workers. Robots did not suffer illness, not even in a pandemic. They did not stop to use the restroom or complain about the speed of the line. Robots never threatened to strike.

A large truck painted purple and white occupied prime position on the showroom floor, a driverless delivery vehicle produced by Gatik, a Silicon Valley company that was running thirty of them between distribution centers and Walmart stores in Texas, Louisiana, and Arkansas. Here was the fix to the difficulties trucking firms faced in attracting and retaining drivers—not better working conditions, but fewer workers.

"It's not quite as appealing a profession as it once was," said the company's head of policy and communications, Richard Steiner. "We're able to offer a solution to that trouble."

An Israeli start-up, SafeMode Mobility, touted a means of limiting the turnover plaguing the trucking industry: an app that monitored the actions of drivers—their speed, the abruptness of their braking, their fuel efficiency—while rewarding those who performed better than their peers.

The company's founder and chief executive officer, Ido Levy, had conceived of the approach with a professor at the MIT Media Lab. They had tapped research on behavioral psychology and gamification.

The SafeMode system was yielding savings of 4 percent on fuel while boosting retention by one-fourth, Levy told me. He displayed data captured the previous day from a driver based in Houston. His steady hand at the wheel had earned him a whopping $8 on top of his usual per mile rate.

"We really convey a success feeling every day," Levy said. "That really encourages retention. We're trying to make them feel that they are part of something."

Such words resonated as a triumph for the consulting class. Technology was being harnessed to *convey a success feeling,* as opposed to delivering demonstrable success via robust pay. Drivers were being made to *feel that they are part of something,* as opposed to actually gaining control over their schedules while capturing some of the profits.

On another section of the showroom floor, major retailers displayed the results of their experiments with drones to deliver packages. One company, Zipline, showed video of its equipment taking off behind a Walmart in Pea Ridge, Arkansas, and then dropping items like mayonnaise and even birthday cakes into the backyards of customer's homes.

Locus, the company that made robots, had already outfitted two hundred warehouses around the world. It had recently expanded into Europe and Australia. Its executives presented the products not as job-killing replacements for human workers but as a complement—a way to squeeze more productivity out of the same warehouse by relieving the humans of the need to push the carts.

There were obvious merits to this framing. Only a Luddite would oppose technology that promised to relieve human beings of dangerous, unpleasant, and grueling work. In the ideal arrangement, robots, drones, and other forms of automation could indeed be entrusted with routinized tasks while allowing humans to be redirected toward more fulfilling pursuits.

The history of automation was full of grave warnings about joblessness that proved hyperbolic. The widespread adoption of automatic teller machines in the 1980s, to cite a famous example, provoked much fretting about potentially jobless bank staff. Yet by 2018, nearly half a million Americans still worked as bank tellers, 10 percent more than in 2000. Because banks were able to operate branches with fewer tellers, they opened more of them, hiring additional people. And they created additional jobs by hiring ATM repair people and other support staff.

There was ample reason to believe that the spread of automation would play out similarly within the supply chain, destroying some jobs while generating more. In countries like Sweden, where labor unions were strong, workers were generally enthusiastic about automation. They understood that if their own roles were taken over by robots, they could count on publicly financed support networks and training to prepare them for their next jobs. Yet in the United States, where social welfare spending was modest, manual workers with fewer specialized skills had legitimate cause for concern. Just as factory work had been dispatched overseas to bypass unions and push down wages, robots and other forms of automation were being embraced by the executive ranks as the latest way to reduce costs and preempt labor agitation while helping themselves to a larger slice of the economic pie.

In the popular narrative, questions about automation and other potent forms of innovation tended to be presented as false binary choices. Were robots good or bad? Would they throw people out of work, or merely supplement human labor, fueling growth that would generate fresh jobs? But this treatment skipped over the most meaningful question. Who got to decide how automation played out, and for whose benefit?

Automation is inevitable for the simple reason that human beings will always apply their creativity to lessen their physical burdens, increase convenience, and expand productivity. What is not inevitable is who will reap the rewards, and whether workers gain a say. In the aftermath of the Great Supply Chain Disruption, the executive class was embracing robots as the solution to supposed worker shortages—in other words, as an alternative to sharing profits by paying higher wages. This was a troubling phenomenon, a sign that those in charge of the supply chain were doubling down on the reckless principles that had brought the system down.

And this was problematic not only for the supply chain, but for human advancement in general. It placed the interests of working people in opposition to technology that held the potential to make work easier. The solution would have been familiar to Ford and Ohno—pay robust wages and restore a sense of security to full-time work, giving employees a proprietary stake in their companies. That would give them an incentive to participate in the reconfiguration of their workplaces. No one wants to be replaced by a robot, but the prospect of controlling one and using it to lighten your load is presumably appealing to all.

But sharing control was not on the menu for those running the businesses at the center of the supply chain. Unions had been decimated over decades, and publicly traded companies had amassed wealth as a result. The robots were being installed at the direction of management, which meant that they were there to serve investors while holding the line on pay for workers. And that meant that consumers would remain exposed to the consequences of workers withholding their labor.

The supply chain leaders convened in Philadelphia grasped the need to indulge the language of collaboration. They took pains to describe automation as a congenial force.

"We have these programs where associates would nickname their bots and get their pictures taken with them," said Nathan Ray, director of distribution center operations at Albertsons, the supermarket chain, who previously held executive roles at Amazon and Target. "It becomes a lot more fun."

But he also acknowledged that robots were a way to "solve labor issues," meaning that companies could take a tough position on pay and minimize their workforces while still moving their products.

Locus explicitly marketed its robots as the solution to labor shortages and the key to maintaining the lean mindset. Unlike human workers, robots could be easily scaled up and cut back, eliminating the need to hire temporary workers, and saving on the time and expense of training recruits.

Party City, a national chain of party supply stores with more than eight hundred outlets across the United States, saw robots as the antidote to paying humans more.

"You couldn't get labor, so you raised your wages to try to get people," said the company's director of transportation, Bruce Dzinski. "And then everybody else raised wages."

In short, the supply chain chaos had restored leverage for workers in their dealings with management—a threat to the bottom line. Companies that were accustomed to treating employees like replaceable parts now felt compelled to increase pay or suffer the consequences of failing to attract and retain enough people.

A labor movement that rarely had cause to celebrate was suddenly tallying substantial victories, not least the establishment of a union at an Amazon warehouse in New York City. Amazon was fiercely opposed to unions, unleashing legions of lawyers and consultants skilled in the arts of suffocating attempts at labor organization. The company's shocking defeat at the hands of a fledgling movement signaled a potentially potent reordering of power.

"It's going to shake up the labor movement and flip the orthodoxy on its head," declared Justine Medina, one of the organizers.

The companies that ran the supply chain were not going to submit to this without a fight. And as I wandered the convention hall in Philadelphia, I could see the outlines of their defense. Workers were mobilizing to capture their piece of the action, threatening to withhold their labor as leverage. So employers were laying preparations for their version of a work stoppage—a hiring strike. They would bolster their own leverage by deploying robots. They would seek to preserve their profits and their privilege by maintaining a permanent state of insecurity within the ranks of the humans they employed.

Just in Case, resilience, sustainability. These were words confined to the talking points that guided supply chain leaders as they described the hard-earned lessons of the pandemic disruptions. But the central motive was unchanged: minimize paychecks.

Which meant that something else was also unchanged: the societal risks of running out of vital goods.

"A GREAT SACRIFICE FOR YOU"

Redrawing the False Bargain

By early 2023, it was reasonable to conclude that the Great Supply Chain Disruption was mostly a spent force.

Shipping rates had plunged back to their historical norms. The American rail strike had been averted, though the dockworkers had yet to reach a deal. Talk of truck driver shortages was largely muted.

In Southern California, Gene Seroka, overseer of the Port of Los Angeles, no longer had to explain why dozens of incoming container ships were stuck at anchor. The floating traffic jams were gone. Instead, he was pressed to account for why cargo volumes were dropping precipitously. In February alone, the number of containers moving through Los Angeles was down 43 percent compared to the previous year.

"The decline was indeed steep," Seroka told a press conference. This, he added, was a "global phenomenon" as the pandemic released its grip on modern life. People were again taking vacations and going out to eat, reducing their demand for physical goods.

In China, extraordinary public protests had produced an abrupt lifting of the Zero COVID policy, allowing factories and ports to resume their usual operations.

The ending of pandemic controls permitted Jacob Rothman—the overseer of the factory in Ningbo—to take a trip outside China to explore other potential factory venues. He visited Walmart's headquarters, touting the merits of his new operations in Southeast Asia. He spent two weeks in Mexico scouting out factory sites. He forged plans to connect Chinese factories with potential Mexican partners.

By many indications, multinational companies were making adjustments to protect their supply chains from the vagaries of geopolitics and trade frictions. Their focus on regional hubs—on nearshoring and reshoring—appeared likely to continue, reducing the world's dependence on Chinese industry. For as long as memories of the pandemic stuck, the dictates of Just in Time would at least be moderated by considerations of Just in Case.

In Europe, the change appeared especially potent. As the European Union sought to limit greenhouse gases reaching the atmosphere, it was extending carbon-reducing regulations to the shipping industry. That would force carriers to purchase so-called offsets—essentially the right to pollute. These rights would cost money, lifting the price of shipping. Which further diminished the merits of using faraway factories to supply Europeans with goods.

And that trajectory was due for a substantial acceleration as another law was phased in beginning in 2024. The European Corporate Sustainability Reporting Directive would require large companies with operations on the continent to publish data disclosing their performance on a range of indicators that tracked their progress in meeting environmental standards. Eventually, they would have to audit their own supply chains and reveal data about their suppliers.

The details were still being crafted, but the new law had already seized the attention of multinational companies along with their law firms, accountants, and consultants. It was being hailed as an agent of transformation throughout the supply chain; the vehicle that would force a transition to environmentally sound pursuits. Surely, that would further degrade the economic logic for shipping factory goods across oceans.

And by the middle of 2023, the world had gained another powerful example of the risks of depending on giant ships to move products around the planet—especially in the face of climate change. The Panama Canal, a conduit for 40 percent of the world's cargo trade, was severely hampered by a terrible drought. The canal required massive infusions of rainwater to maintain flow. With the skies withholding rain, the canal authority was restricting traffic, while cargo operators themselves were limiting freight to avoid hitting bottom. Scores of vessels were stuck floating at both ends of the canal, bobbing in the Atlantic and the Pacific—a warning sign of shifts that were certain to intensify. That added to the momentum for reshoring and nearshoring.

And then came still another reminder of the perils of entrusting the global supply chain to container vessels. In December 2023, Iranian-backed rebels in Yemen unleashed drone and missile strikes on ships passing through the Red Sea, a busy entry point to the Suez Canal for vessels headed to Europe from Asia. This was a show of solidarity with Palestinians under sustained bombing in Gaza. Many carriers began bypassing the canal by embarking on far longer, costlier journeys around the African continent. The resulting delays and product shortages were another reason to contemplate the merits of nearshoring.

In areas of the American economy where deregulation and market concentration had given businesses the power to manipulate the supply of goods to lift prices, a new spirit of government oversight was palpable.

A bill working its way through Congress in the spring of 2023 aimed to strip the shipping industry of its traditional exemption from American antitrust law. Other proposed legislation sought to force vessels to load agricultural exports—an effort to protect farmers like Scott Phippen from being denied containers by ocean carriers that were cashing in on surges of imports. As the Federal Maritime

Commission implemented the new shipping act in the summer of 2023, it was writing rules aimed at ensuring that agricultural exports would find space on deck.

Coy Young, the Missouri rancher who nearly took his life after his disastrous cattle sale, testified before a congressional panel that was probing price fixing in the beef industry—a hearing prompted by my reporting. The Department of Agriculture began writing new rules aimed at diminishing the power of the meatpackers by reinvigorating prohibitions against market manipulation in the Packers and Stockyards Act.

A reactivated Federal Trade Commission was reviving antitrust law, subjecting corporate mergers to scrutiny while threatening to break up the giants.

Biden had gratified the railroads in their opposition to paid sick leave, imposing the deal that averted the strike with only a single personal day. But he used the bully pulpit to pressure railroads to individually extend sick leave. By late March 2023, Union Pacific had struck agreements with eight unions to extend up to seven days of such leave per year.

The following month, Biden's national security advisor, Jake Sullivan, effectively declared last rites on the philosophy that had propelled globalization for decades.

"In the name of oversimplified market efficiency, entire supply chains of strategic goods—along with the industries and jobs that made them—moved overseas," Sullivan said. "And the postulate that deep trade liberalization would help America export goods, not jobs and capacity, was a promise made but not kept."

The corrective, he asserted, was "a modern American industrial strategy" in which the government would reorient trade policy and invest unabashedly toward boosting manufacturing capacity at home.

There were worrying signs that an obsession with containing China had reached feverish proportions in Washington, shunting aside all other objectives. Indonesia held the world's largest reserves of nickel, a critical mineral used to make electric vehicle batteries. The Indonesian government was eager for a trade deal with the United States that would bring American investment and technology to exploit the country's nickel while unlocking the tax credits Biden had crafted to encourage the building of electric vehicle factories. But the Biden administration was rejecting those entreaties because Indonesia was heavily dependent on Chinese investment for processing nickel. The imperatives to adapt to climate change and configure a resilient supply chain took a back seat to isolating those willing to do business with Chinese companies.

Still, taken broadly, the fresh policies prevailing in Washington indicated a powerful shift in the political firmament. They affirmed the widening view that shortages of goods, rising prices for consumer wares, and deep-seated anxieties about globalization were in part the result of excessive deregulation combined with reckless faith in a fetishized notion of efficiency.

The industries at the center of the supply chain—from shipping and rail to meat processing—had liberated themselves from rules imposed to limit their dominance. They had reprised the era of the Robber Barons in achieving monopoly status. This had delivered stupendous profits to shareholders while yielding danger and dysfunction for society at large.

The chaos at ports, the truck driver shortages, the train accidents, and the COVID-19 deaths inside slaughterhouses were all testaments to the consequences of this bargain. The reliability of American transportation had been downgraded in exchange for fatter dividends for investors. Here was a transfer of wealth from the bottom up, from the masses to the fortunate few.

And now the government was seeking redress, though in patchwork fashion, and with no guarantee of success or follow-through.

In the arc of American history, the monopolists had enjoyed quite a run. They would not surrender their power this time—or anytime—without a fight. They were unleashing their lobbyists to defang the campaign to reinstall the sorts of regulations that had worked to protect the economy from extreme market concentration in another era.

Still, the consensus that had captured mainstream economics and both major political parties for more than half a century—the notion that scale and efficiency were good for American consumers—had given way to a debate that included healthy skepticism and an appreciation for the lessons of history. A battle was underway over the future of the supply chain. Corporate mergers were again facing scrutiny. Labor was projecting newfound power in pursuit of higher wages.

And the contours of international commerce were again being reconfigured—this time with an emphasis on regional hubs for manufacturing. It was conceivable that the next time Hagan Walker went looking for a factory to make his latest products, he might discover options closer to home.

This was one possible outcome—that the supply chain would be reshaped with greater value placed on resilience—but it was no sure thing. The focus on nearshoring and reshoring could yet be reversed, and for a simple reason: because a system of global manufacturing and retail dominated by companies that answered to Wall Street would have limited willingness to pay extra to make its products. Spending more as a hedge against risks that were eventually inevitable yet hypothetical in any given quarter was a good way for an executive to lose their job. That essential reality had not changed.

The share of American imports that came from China had dropped by 5 percent between 2017 and 2022. Many of the goods that used to be manufactured inside Chinese factories were instead being made in Vietnam, where they cost 10 percent more on average, according to one study. Orders that had shifted to Mexico were 3 percent more expensive compared to those manufactured in China.

Perhaps, like Columbia Sportswear, more brands would prove willing to pay those higher prices as a way to cut the distance between their factories and their markets, while focusing on the savings derived from avoiding tariffs or limiting shipping costs. Maybe they would justify higher costs as a means of reducing their carbon emissions as they sought to win favor with consumers concerned about climate change. The brands might conceivably accept higher prices as a way to unlock greater savings through leaner inventories, gaining the ability to replenish their stocks more quickly. All of this was possible and even hopeful, and still the signs were emerging that the old normal was making a comeback.

As multinational companies shifted factory orders to other countries, the price of making clothing in China was dropping precipitously, which was prompting some brands to reverse course. "China is getting so cheap that a lot of companies are going back," Bernardo Samper, a longtime sourcing agent in New York, told me in October 2022. "At the end of the day, everything is driven by pricing."

Chinese apparel makers that had set up factories in Vietnam and Cambodia to get around American tariffs were frustrated by their inability to secure parts and materials. Some were moving back to China.

In Guatemala City, even the head of the family-owned factory that was aiming to secure extra business from Columbia Sportswear wondered if Latin America could ever really compete with businesses on the other side of the Pacific.

"People tend to gravitate toward lower prices in Asia," Juan A. Sanchez, chief executive of Zuntex, told me. "Nobody gets fired for going to lower prices."

It was hard to argue with his logic. It was equally hard to shake the sense that unless the world finds a different guiding principle for how to make and deliver the products of modern life, the whole system is certain to break down again.

The journalist Barbara Ehrenreich once described the exertions of poorly paid employees as a universal subsidy that held down the cost of living for all.

"When someone works for less pay than she can live on—when, for example, she goes hungry so that you can eat more cheaply and conveniently—then she has made a great sacrifice for you, she has made you a gift of some part of her abilities, her health, and her life," Ehrenreich wrote in her masterful book *Nickel and Dimed*. "The 'working poor,' as they are approvingly termed, are in fact the major philanthropists of our society. They neglect their own children so that the children of others will be cared for; they live in substandard housing so that other homes will be shiny and perfect; they endure privation so that inflation will be low and stock prices high. To be a member of the working poor is to be an anonymous donor, a nameless benefactor, to everyone else."

Those words were powerful and apt, describing the conditions on which normalcy rested. As we have seen, a functioning supply chain has come to depend on routine forms of exploitation. Keeping store shelves stocked has in recent decades demanded that truck drivers miss their wedding anniversaries. Maintaining the flow of cargo has required that rail workers remain on the road for weeks at a time, even when their children need surgery. The sanctity of the beef supply seemingly mandated that Tin Aye would never hold her only grandchild.

These are false bargains. They rest on the assumption that corporate executives must gain the largess to amass private islands or the whole system breaks down. We can operate the supply chain, producing and delivering the wares of the modern world, and still protect and justly compensate ordinary workers. The tragic trade-offs make sense only so long as we continue to be guided by the fatuous assumption that shareholders and consumers are fundamentally conjoined in their interests, with workers posing a threat to both.

That assumption still has life.

As inflation cooled throughout 2023—in part because supply chains were regaining functionality—the American unemployment rate began edging up. This seemed to affirm the central bank's policy: Higher interest rates were slowing economic activity, which spelled less need for workers. Every month that joblessness went up, stock markets rallied. The direct cause of investor joy was the indication that interest-rate increases were nearing their end, allowing greater borrowing to resume—a boon for stocks. But the relationship between worker distress and stock market happiness was also telling a deeper story. Here was a clear sign that the people in control of money craved the low wages that came from a permanent state of insecurity among working people. The economy was still ruled by the idea that if employees took home too much, employers could not prosper, which meant that the supply chain and the broader economy remained at risk.

The reconfiguration of the global supply chain that is underway may be able to adjust to the geopolitical alterations, but it will not reckon with this fundamental vulnerability—the permanent fragility arising from dependence on exploited labor.

We cannot count on a transportation system staffed by people who must perpetually choose between protecting their health or their jobs. When we downgrade work to the point that sacrificing one's basic needs becomes a job requirement, we confront the eternal risk that some will withhold their labor. When we allow monopolies to capture our markets, scarcity and rising prices are the inevitable result.

It took a rare event—literally a once-in-a-lifetime pandemic—to render these truths undeniable. But the uniqueness of that perspective does not make the lessons any less true.

We cannot predict the details or timing of the next shock to the system, but we know that it will come. And when it does, we are likely to be here again, watching the supply chain buckle, our productive capacities falling prey to our failure to ensure that the people doing the work are motivated by the ultimate incentive: a fair deal.

In an era of extraordinary economic inequality, the supply chain is that rare thing that holds universal import. We all depend on it, even as—most of the time—we need not give a thought to how it operates or who is keeping it going. But its breakdown demands reflection on what happened, and what must happen next.

When the wealthiest country on earth runs out of protective gear in the middle of a public health catastrophe, our resources and know-how have clearly been sabotaged by financial interests. When parents cannot locate crucially needed infant formula, we justifiably surrender faith in the workings of the modern marketplace. When enormous companies spend billions of dollars making cars that no one can drive for a lack of chips, our system has failed. And when the limited supply of chips is seized by the makers of smartphones to the exclusion of lifesaving medical devices, we can only conclude that we have surrendered something elemental in the relentless pursuit of profit—our status as a civilized society.

That many of these disasters stem from the overly aggressive deployment of Just in Time must go down as a bitter historical irony. The Toyota production system was developed as an innovative way to limit waste. The version of *lean* that has prevailed across global business—a trend inspired by Toyota, yet hijacked by consultants—has produced extraordinary waste, from the Ford trucks idled next to Henry Ford Elementary School to the mountains of almonds stuck in California because of a shortage of containers.

The same consultants who constructed the Just in Time supply chain are now marketing technological solutions to the delicate world they created—apps that allow greater transparency in shipping, tracking devices, software systems that can detect bottlenecks and work around them, online platforms that coordinate factory production on multiple shores.

Technology that brings together now-disparate pieces of the supply chain is full of possibilities for improvement. Surely, some of these creations will prove useful, but none can relieve the global economy from its exposure to the vulnerabilities arising from the ultimate threat to the supply chain: unregulated greed.

The buzzwords will change along with the geography of international trade. Inventory will fluctuate depending on our proximity to crisis. But the same pitfalls will remain, awaiting the next rupture.

The ultimate solution to the upheaval in the supply chain is the resumption of sensible regulations that promote competition and allow workers to bargain for their share of the proceeds. We need rules that preserve the innovation and growth that flow from capitalism while also protecting society from the clearly established tendency toward monopoly. We need real and transparent marketplaces that are fair to all participants while maximizing the powers of supply and demand—for commodities, for rail and shipping services, for labor.

This is not a utopian vision, but rather a return to the mode of governance that prevailed in the United States from the end of World War II through the late 1970s. We do not want a time machine back to that era. We can hang on to our technological gains and our social progress while still resurrecting crucial policies that were dismantled over decades by monopolists in pursuit of fatter profits. Antitrust enforcement is required to ensure real efficiency, the kind that flows from a truly competitive marketplace. Transparency in markets, from transportation to agriculture to labor, is needed to prevent the largest players from abusing their dominance to the detriment of everyone else.

The journey of Hagan Walker's container from Ningbo to Starkville has brought home one fact above all. The contemporary supply chain was constructed by and for the benefit of enormous publicly traded companies, from big-box retailers like Walmart to e-commerce giants like Amazon along with sprawling ocean carriers, railroad monopolists, and agribusiness goliaths. They have justified their dominance on the strength of what they have delivered to consumers: low prices. But the pandemic has revealed the full costs of the bargain—the shortages of vital products, the flimsiness of transportation systems that depend on exploited workers, the horrible sacrifices of vulnerable people like Tin Aye.

We have surrendered our basic decency along with our sense of economic security.

This is how the world ran out of vital creations at the very worst time. The people in charge of the supply chain built it to ensure that their profits would never be scarce. It is time for broader interests to gain a say over how we receive the goods that land at our doors.

ACKNOWLEDGMENTS

As often happens in journalism, the discovery that turned into this book was a lucky accident. I was living in London, working on yet another story about Britain's ill-fated abandonment of the European Union, when an electronics importer suggested that I investigate the chaos at the ports. The price of shipping goods from Asia to the United Kingdom was skyrocketing. This turned out to be true, but it was merely part of a far larger story. Around the globe, the shipping industry was going haywire. I did some digging and learned about what I came to call the Great Supply Chain Disruption. Then I spent the next two years tracing its roots and exploring its consequences, from Asia to the Middle East to North America.

That such an undertaking was even imaginable speaks to my enduring fortune in working for the *New York Times*. No newsroom on earth is more committed to devoting the time and resources needed to bring complex and important projects to fruition. Many thanks to the ultimate guardians of the faith in the value of reporting, A. G. Sulzberger and Joe Kahn. I was blessed to collaborate with one of the supreme editors in the business, Rich Barbieri, who, among other things, concocted the genius headline that I have shamelessly stolen as the title for this book.

The Business desk has been my journalistic home for nearly two decades (not counting an unexpected hiatus for a few years). I am grateful for the support and extraordinary humanity of Business editor Ellen Pollock, an indefatigable advocate for a good yarn. I am thankful for the sharp eyes, collegiality, and wisdom of other colleagues on the desk—Kevin McKenna, Vikas Bajaj, Ashwin Seshagiri, David Enrich, Rachel Dry, Phyllis Messinger, Dave Schmidt, Kevin Granville, Howell Murray, and Renee Melides.

Much gratitude to fellow writers and data-visual wizards I teamed up with and learned from along the way—Alexandra Stevenson, Keith Bradsher, Vivian Yee, Ana Swanson, Patricia Cohen, Lazaro Gamio, Vivian Wang, Eshe Nelson, Jack Ewing, Steve Lohr, and Noam Scheiber.

That the *Times* remains unwavering in its commitment to serious international journalism attests to the character and dedication of the people running the place. Many thanks to Phil Pan, Adrienne Carter, Michael Slackman, Jim Yardley, Kim Fararo, Claire Gutierrez, Greg Winter, Matt Purdy, and Carolyn Ryan.

Among the continued joys of working at the *Times* is participating in a constantly evolving range of audio projects. Some of the elements in this book were aired out in part through episodes of *The Daily,* under the brilliant guidance of Lisa Tobin, Paige Cowett, and Diana Nguyen, and via the disarmingly thoughtful interrogation of Michael Barbaro, Sabrina Tavernise, and Natalie Kitroeff. Many thanks to Robert Jimison, Lynsea Garrison, John Ketchum, and Michael Johnson.

I am grateful for the time and intelligence of several people who read versions of the book and offered counsel. My great friend, the world-class writer David Segal, advocated an especially useful change. Many thanks to Suzanne Berger at MIT and Ambassador Derek Shearer at Occidental College. David Sirota—whose digital news platform, *The Lever,* has become an indispensable read for those exploring the intersection of money and politics—is a never-ending fountain of clarifying insights and sharp questions that have found their way into this book.

I am uniquely indebted to Adil Ashiq at AIS Maritime Intelligence and William George at ImportGenius, who patiently tore through reams of data in allowing me to x-ray the epic traffic jams off the Southern California ports. And special thanks to Shelly Dekalo and Freightos, for putting me in touch with Hagan Walker and Glo. Raine Mahdi at Zipfox put me on the path toward Laredo, Texas, and Mexico.

One key idea at the center of this book—that shortages of goods were in large part the outgrowth of excessive dependence on Just in Time manufacturing—crystallized in a conversation with the economist Ian Goldin at Oxford University. Chad Bown, resident trade genius at the Peterson Institute for International Economics in Washington, has guided me through enough supply chain questions to fill the *Maersk Emden*. Willy Shih at Harvard Business School was generous with his time, case studies, and wisdom gleaned from decades of travel to industrial zones around the world. Pietra Rivoli at Georgetown University's McDonough School of Business first schooled me in the complexities of the global supply chain.

A big thank-you to the Camden Conference in Maine for convening an especially pertinent and provocative conversation about the questions that fill out part III of this book, affording me the chance to sharpen my thinking with key experts on international trade— Jennifer Hillman at Georgetown; Caroline Freund at the University of California, San Diego; David Autor at MIT; Simon Evenett at the University of St. Gallen in Switzerland; Mark Wu at Harvard Law School; Douglas Irwin at Dartmouth; Ajay Chhibber at George Washington University; and former US senator John E. Sununu. And thanks to public-radio legend David Brancaccio, for his engaging turn as maestro.

Much appreciation for my decades of conversation with the legendary economist Joseph Stiglitz, whose remarkable books on globalization, international trade, and economic inequality have influenced my thinking on all of these subjects. Many thanks to Rob Johnson at the Institute for New Economic Thinking and William Lazonick at the University of Massachusetts, for crucial contributions to my understanding of financialization in shaping the workings of global supply chains.

In Detroit, Mike Skinner guided me through the Ford Piquette Avenue factory and shared archives of Ford-related biographical material. Matt Anderson, curator of transportation at the Henry Ford Museum of American Innovation in Dearborn, Michigan, led me on a useful tour.

On the subject of how the consulting class warped the Toyota production system, I'm grateful for the generosity of my colleague Michael Forsythe, whose own book (with Walt Bogdanich) *When McKinsey Comes to Town: The Hidden Influence of the World's Most Powerful Consulting Firm* is a must-read.

Any writer who wades into the subject of container shipping owes a debt to the economic historian Marc Levinson and his masterwork, *The Box: How the Shipping Container Made the World Smaller and the World Economy Bigger*. Many thanks to Alan Murphy at Sea-Intelligence in Copenhagen and Florian Frese at Container Xchange in Hamburg for helping decode the constantly refashioning puzzle of global freight patterns. Much respect to *FreightWaves* and senior editor Greg Miller for probing coverage of the container industry. Lorela Sandoval in Manila helped track down container ship crew members.

In exploring China's entry into the global trading system, I drew on the work of my former *Washington Post* colleagues John Pomfret and Paul Blustein, journalist Jim Mann (an early and inveterate skeptic of "constructive engagement"), and Nicolas Lardy at the Peterson Institute.

Many thanks to Robert Morris, who went well beyond the call of duty as communications officer for the Port of Savannah in securing a boat for an up-close look at the floating traffic jam. And thanks to photographer Erin Schaff, for a fruitful collaboration. Ditto to Phillip Sanfield at the Port of Los Angeles.

Thanks to Steve Wen at Dray Alliance in Long Beach for securing my ride alongside a dray operator. And thanks to David Martin in Chattanooga and Max Farrell at WorkHound, for helping me gain entry to the passenger seat of a long-haul truck. Here's to photographer George Etheredge, for our frozen adventure across the midsection of the United States. I appreciate the time, experience, and stubborn truth telling of Steve Viscelli, the trucking expert at the University of Pennsylvania.

In probing the subject of monopoly power, I am grateful for the work of Matt Stoller at the American Economic Liberties Project, David Dayen and Matthew Jinoo Buck at the *American Prospect*, and Barry Lynn at the Open Markets Institute.

Thanks to Dustin Ogdin at the Northern Plains Resource Council, for putting me in touch with Montana cattle ranchers.

In examining the future of globalization, I was influenced by two excellent and timely books, Shannon O'Neil's *The Globalization Myth: Why Regions Matter* and Rana Foroohar's *Homecoming: The Path to Prosperity in a Post-Global World*. Many thanks to Keith Rockwell, a global fellow at the Wilson Center, for never treating stupid questions about trade like stupid questions.

Gail Ross is the greatest literary agent in the business—whip-smart, appropriately skeptical, yet full of enthusiasm and ambition. Peter Hubbard, my editor at HarperCollins, is a writer's dream. Many thanks to his world-beating team—Jessica Vestuto, Maureen Cole, Kayleigh George, and Molly Gendell.

Shout-outs to my community of wonderful friends in the journalism world and beyond, for support that was especially critical this time around—Hannah Beech and Brook Larmer, Joshua Shulman, Ken Belson, Gady Epstein, Barbara Demick, Jesse Eisinger, Jesse Drucker, Michael Powell, Larry and Vicki Ingrassia, and Chris Rose.

If there is a way to write a book without imposing heavily on one's family, I haven't found it yet. Forever thanks to my in-laws for keeping the smaller people fully stocked with love, amusement, and sustenance during my too-frequent absences—Mimi and Donald Fei, Jessica Fei, and Geobany Rodriguez. Thanks to my mother, Elise Goodman; my sister, Emily Goodman;and her husband, Joe Kaufman. Thanks to everyone in Croton who has ever cooked us a meal or given one of my kids a ride when I've not been around to do it myself.

The words are all too puny to convey my appreciation for my immediate family. My daughter Leah Goodman is far away in physical space but forever close at heart. My son Leo Fei-Goodman dazzles me with his probing questions, his never-ending goodness, and—lately—his entrepreneurial inclinations. My daughter Mila Fei-Goodman brings the soul of a writer into everything, her keen powers of observation a perpetual inspiration. Our youngest, Luca Fei-Goodman, constantly reminds me of the sheer wonder in watching heavy equipment at work. ("Digger! Digger!") This book was written very much despite him, and yet because of him at the same time. My wife, Deanna Fei, is the center of all that is true and meaningful in my life. She generously set aside her own work to administer a rigorous edit, grappled with attending to three kids solo while I was on the road, and never failed to provoke that tingly feeling whenever she walked into the room. This book is for her.

NOTES

PROLOGUE: "THE WORLD HAS FALLEN APART."

2 *138 containers:* Data furnished by Adil Ashiq at AIS Marine Intelligence, provided by MarineTraffic.

2 *13 million pounds:* Ibid.

2 *More than 17 million pounds:* Ibid.

2 *$25 billion worth of goods:* Greg Miller, "$25B Worth of Cargo Stuck on 80 Container Ships off California," *American Shipper,* October 20, 2021.

2 *Nearly 13 percent:* Data furnished by Sea-Intelligence.

3 *1,200 feet long: Maersk Emden* description on VesselFinder, https://www.vessel finder.com/vessels/MAERSK-EMDEN-IMO-9456769-MMSI-219056000.

3 *roughly twelve thousand containers:* Data analysis by ImportGenius.

3 *474 containers for LG:* Analysis of US Customs disclosures furnished by William George at ImportGenius.

6 *plunged to a record low:* Auto Inventory/Sales Ratio, FRED Economic Data, St. Louis Fed, https://fred.stlouisfed.org/series/AISRSA.

CHAPTER 1: "JUST GET THIS MADE IN CHINA."

24 *"We can't continue to allow China":* Trump at rally in Fort Wayne, Indiana, May 1, 2016, https://www.cnn.com/videos/politics/2016/05/01/donald-trump-china -rape-our-country.cnn.

25 *30 million people:* John Pomfret, *The Beautiful Country and the Middle Kingdom: America and China, 1776 to the Present* (New York: Henry Holt, 2016), chapter 29.

25 *killed as many as 8 million:* Yongyi Song, "Chronology of Mass Killings during the Chinese Cultural Revolution (1966–1976)," Sciences Po, August 25, 2011, https://

www.sciencespo.fr/mass-violence-war-massacre-resistance/en/document
/chronology-mass-killings-during-chinese-cultural-revolution-1966-1976
.html.

26 *Foreign investment:* Wenhui Fan, "Foreign Direct Investment in China: 1981–
2001," East-West Center Working Paper No. 30, 2006, 3.

26 *making motorcycles:* Paul Blustein, *Schism: China, America and the Fracturing of the
Global Trading System* (Waterloo, Ont.: Centre for International Governance Innovation, 2019), 23.

26 *saw its exports:* Joe Studwell, *The China Dream: The Quest for the Last Great Un-
tapped Market on Earth* (London: Profile Books, 2002), 45.

26 *nearly 10 percent:* Edward S. Steinfeld, *Forging Reform in China: The Fate of State-
Owned Industry* (Cambridge, UK: Cambridge University Press, 1998), 1.

26 *In Shenzhen:* Peter S. Goodman, "In China, Building Worries; As Housing
Keeps Going Up, Some Fear the Bubble Will Burst," *Washington Post,* March 5,
2003, E1.

26 *artificial Christmas trees:* Peter S. Goodman, "Capitalizing on Christmas; America's Celebration Is China's Windfall," *Washington Post,* November 9, 2003, F1.

26 *15 million microwave ovens:* Peter S. Goodman, "China Is Resisting Pressure to
Relax Rate for Currency," *Washington Post,* September 1, 2003, A1.

26 *Deng's economic reforms:* "Four Decades of Poverty Reduction in China: Drivers,
Insights for the World, and the Way Ahead," World Bank Group, Development
Research Center of the State Council, the People's Republic of China, 2022, 2.

28 *"butchers of Beijing":* "Clinton Campaign Asks Who's the Real Foreign Policy
Risk," U.S. Newswire, October 12, 1992.

28 *"what those kids did":* Michael Wines, "Bush, This Time in Election Year, Vetoes
Trade Curbs Against China," *New York Times,* September 29, 1992, A1.

28 *"Observe human rights":* "Campaign '92: Transcript of the First Presidential Debate," October 12, 1992, *Washington Post,* A16.

28 *he signed an executive order:* Blustein, *Schism,* 33.

29 *"The American people admire":* Paul Blustein, "Little-Stick Diplomacy; Clinton,
Jiang Find Harmony," *Washington Post,* June 29, 1998, B1.

29 *shark's fin and grilled beefsteak:* Ibid.

29 *conducted the band:* Ibid.

30 *urban Chinese workers:* Steinfeld, *Forging Reform in China,* 16.

30 *outstanding loans:* Nicholas R. Lardy, *Integrating China into the Global Economy*
(Washington: Brookings Institution Press, 2002), 20.

30 *state-owned companies:* "2004 Report to Congress on China's WTO Compliance,"
United States Trade Representative, December 11, 2004, 3.

30 *financial giants:,* Lardy, *Integrating China,* 2–4.

31 *"not simply agreeing":* President Clinton's speech on China trade bill, delivered at
Paul H. Nitze School of Advanced International Studies, Johns Hopkins University, in Washington, March 8, 2000, as recorded by Federal News Service.

32 *"In the new century":* Ibid.

32 *"sow the seeds of freedom":* Pomfret, *The Beautiful Country,* 44.

32 *hostilities over Taiwan:* Jennifer A. Hillman, "China's Entry into the WTO—A Mistake by the United States?" Georgetown University Law Center, 2022, 5–6.

32 *Major lobbying organizations:* Robert Dreyfuss, "The New China Lobby," *American Prospect,* December 19, 2001, https://prospect.org/world/new-china-lobby/.

33 *"convinced the president":* Hillman, "China's Entry into the WTO," 7.

33 *beyond $50 billion a year:* Fan, "Foreign Direct Investment," 3.

33 *$14 billion in investment:* Marc Levinson, *The Box: How the Shipping Container Made the World Smaller and the World Economy Bigger* (Princeton, NJ: Princeton University Press, 2016), 362.

33 *worth $132 billion:* Jude Blanchette, Jonathan E. Hillman, Maesea McCalpin, and Minda Qiu, "Hidden Harbors: China's State-Backed Shipping Industry," *CSIS Briefs,* Center for Strategic & International Studies, July 2020.

34 *on Hong Kong:* Chris Buckley, Vivian Wang, and Austin Ramzy, "Crossing the Red Line: Behind China's Takeover of Hong Kong," *New York Times,* June 28, 2021, A1.

34 *from $272 billion in 2001:* World Bank data, Exports of Goods and Services (current US$)—China.

34 *the world's air conditioners:* "Made in China?" *The Economist,* March 12, 2015.

34 *the world's solar panels:* Henry Wu, "The United States Can't Afford the Brutal Price of Chinese Solar Panels," *Foreign Policy,* July 14, 2021.

34 *Not even pharmaceutical companies:* Peter S. Goodman, "China's Killer Headache: Fake Pharmaceuticals," *Washington Post,* August 30, 2002.

35 *American manufacturing jobs:* David H. Autor, David Dorn, and Gordon H. Hanson, "The China Shock: Learning from Labor Market Adjustment to Large Changes in Trade," National Bureau of Economic Research Working Paper No. 21906, January 2016.

35 *"a retailer in America":* House Ways and Means Committee, Subcommittee on Trade, Hearing on China's Most-Favored Nation Status, June 17, 1998, transcript by Federal Document Clearing House.

36 *boosted spending power:* Xavier Jaravel and Erick Sager, "What Are the Price Effects of Trade? Evidence from the U.S. and Implications for Quantitative Trade Models," Centre for Economic Performance, London School of Economics, CEP Discussion Paper No. 1642, August 2019, 5, n. 6, https://cep.lse.ac.uk/pubs/download/dp1642.pdf.

36 *between 2004 and 2015:* Liang Bai and Sebastian Stumpner, "Estimating U.S. Consumer Gains from Chinese Imports," *American Economic Review: Insights* 1, no. 2 (September 2019): 209–24, https://www.aeaweb.org/articles?id=10.1257/aeri.20180358.

36 *factories across China:* Peter S. Goodman and Philip P. Pan, "Chinese Workers Pay for Wal-Mart's Low Prices," *Washington Post,* February 8, 2004.

37 *80 percent were in China:* Ibid.

37 *spending $15 billion:* Ibid.

37 *beyond $200 billion:* Devon Pendelteon and *Bloomberg,* "World's Richest Family Loses $19 Billion on Walmart's Biggest One-Day Wipeout Since 1987," *Fortune,* May 18, 2022.

37   *fewer than one-third:* Andreas Oeschger, "How to Recalibrate Trade Adjustment Assistance to Help Workers Hurt by Trade Liberalization," International Institute for Sustainable Development, September 26, 2022, https://www.iisd.org/articles/policy-analysis/trade-adjustment-assistance-help-workers.

37   *"overcorrection in the narrative":* Interview with Jessica Chen Weiss, October 19, 2023.

CHAPTER 2: **"EVERYONE IS COMPETING FOR A SUPPLY LOCATED IN A SINGLE COUNTRY."**

40   *covered up reports:* Jane Li, "Martian Language, Emoji, and Braille: How China Is Rallying to Save a Coronavirus Story Online," *Quartz,* March 11, 2020.

41   *isolating Wuhan:* Amy Qin and Vivian Wang, "Wuhan, Center of Coronavirus Outbreak, Is Being Cut Off by Chinese Authorities," *New York Times,* January 23, 2020, A1.

41   *forcibly rounding up people:* Amy Qin, "China Expands Chaotic Dragnet in Coronavirus Crackdown," *New York Times,* February 14, 2020, A1.

41   *300 million migrant workers:* Javier C. Hernandez, "Coronavirus Lockdowns Torment an Army of Poor Migrant Workers in China," *New York Times,* February 24, 2020, A10.

41   *workers and parts:* Keith Bradsher, "Slowed by the Coronavirus, China Inc. Struggles to Reopen," *New York Times,* February 17, 2020, A1.

41   *plunged by 17 percent:* "China's Exports Plunge Amid Coronavirus Epidemic," *Wall Street Journal,* March 7, 2020.

41   *average of thirty thousand:* Yossi Sheffi, *The Magic Conveyor Belt: Supply Chains, A.I., and the Future of Work,* (Cambridge, MA: MIT CTL Media, 2023), 2.

42   *forty-three different countries:* Magdalena Petrova, "We Traced What It Takes to Make an iPhone," CNBC, December 14, 2018, https://www.cnbc.com/2018/12/13/inside-apple-iphone-where-parts-and-materials-come-from.html.

42   *products like disposable diapers:* Sheffi, *The Magic Conveyor Belt,* 5.

42   *fifty-one thousand companies worldwide:* Marc Levinson, *Outside the Box: How Globalization Changed from Moving Stuff to Spreading Ideas* (Princeton, NJ: Princeton University Press, 2020), 227.

42   *it would miss targets:* Daisuke Wakabayashi, "Apple Signals Coronavirus's Threat to Global Business," *New York Times,* February 18, 2020, A1.

42   *a dearth of parts:* Ben Foldy, "Coronavirus Fallout Threatens Auto Industry's Supply Chain," *Wall Street Journal,* February 7, 2020.

42   *Fiat Chrysler:* Ibid.

42   *The French automaker:* Ibid.

42   *So did Hyundai:* Jack Ewing, Neal E. Boudette, and Geneva Abdul, "Virus Exposes Cracks in Carmakers' Chinese Supply Chains," *New York Times,* February 5, 2020, B1.

42   *Fashion companies scrambled:* "FACTBOX—Companies Feel Impact of Coronavirus Outbreak in China," Reuters, February 12, 2020, https://www.reuters.com/article/uk-china-health-business-impact-factbox/factbox-companies-feel-impact-of-coronavirus-outbreak-in-china-idUKKBN1ZZ0AV.

42 *85 percent of global production:* Paul Ziobro, "Coronavirus Upends Global Toy Industry," *Wall Street Journal,* February 27, 2020.

42 *Nintendo delayed deliveries:* Peter S. Goodman, "A Global Outbreak Is Fueling the Backlash to Globalization," *New York Times,* March 7, 2020, B1.

42 *imported electronic components:* Ibid.

42 *A lingerie company:* Jon Emont and Chuin-Wei Yap, "Companies That Got Out of China Before Coronavirus Are Still Tangled in Its Supply Chains," *Wall Street Journal,* March 8, 2020.

43 *giant rolls of paper:* Michael Corkery and Sapna Maheshwari, "Is There Really a Toilet Paper Shortage?" *New York Times,* March 13, 2020.

43 *enough protective gear:* Melanie Evans and Drew Hinshaw, "Masks Run Short as Coronavirus Spreads; Hospitals in Europe and the U.S. Hunt for Medical Supplies to Treat Coronavirus Cases," *Wall Street Journal,* February 27, 2020.

43 *hospitals and nursing homes:* Ibid.

43 *telling hospitals:* Ibid.

43 *thousands of elderly people:* Peter S. Goodman and Erik Augustin Palm, "Pandemic Exposes Holes in Sweden's Generous Social Welfare State," *New York Times,* October 9, 2020, B1.

43 *ventilators from China:* "COVID-19: China Medical Supply Chains and Broader Trade Issues," Congressional Research Service, R46304, 33.

43 *many antibiotics:* Ana Swanson, "Virus Spurs U.S. Efforts to End China's Chokehold on Drugs," *New York Times,* March 12, 2020, B3.

43 *suppliers of basic chemicals:* Peter S. Goodman, Katie Thomas, Sui-Lee Wee, and Jeffrey Gettleman, "A New Front for Nationalism: The Global Battle Against a Virus," *New York Times,* April 12, 2020, BU1.

44 *export bans:* Ibid.

44 *American hospitals:* Knvul Sheikh, "Essential Drug Supplies for Virus Patients Are Running Low," *New York Times,* April 2, 2020.

44 *"competing for a supply":* Interview with Rosemary Gibson, April 6, 2020.

46 *"he could be pinched":* Steven Watts, *The People's Tycoon: Henry Ford and the American Century* (New York: Vintage Books, 2006), chapter 14.

46 *two thousand acres of:* Ibid.

46 *"some of every part":* Henry Ford, introduction to *My Life and Work* (1922; repr., Pantianos Classics, ebook).

46 *"a vast extension":* Ibid.

47 *rainforest of the Amazon:* Greg Grandin, introduction to *Fordlandia: The Rise and Fall of Henry Ford's Forgotten Jungle City* (New York: Henry Holt, 2009).

47 *"The trouble with us today":* "Change Is Not Always Progress," in *Ford Ideals: Being a Selection from "Mr. Ford's Page" in The Dearborn Independent* (Dearborn, MI: Dearborn Publishing, 1922), 358–59; cited in Watts, *The People's Tycoon,* chapter 20.

47 *tried to unionize:* Watts, *The People's Tycoon,* chapter 22.

48 *supposed to be a farmer:* Ibid., chapter 1.

48 *encountered a spectacle:* Ibid.

48 *"mostly of the difficulties":* Henry Ford and Samuel Crowther, "The Greatest American," *Cosmopolitan* (July 1930): 38–39; cited in Watts, *The People's Tycoon,* chapter 3.

49   *"for the great multitude"*: Ford, *My Life and Work,* chapter 4.

49   *stock control department:* Watts, *The People's Tycoon,* chapter 7.

50   *more than doubled:* Ibid., chapter 8.

50   *in the United States:* Ibid.

50   *a mere ninety-three minutes:* Ibid.

50   *from $950 in 1910:* Richard Snow, *I Invented the Modern Age: The Rise of Henry Ford* (New York: Scribner, 2013), 214.

50   *"buying materials"*: Ford, *My Life and Work,* chapter 10.

50   *"largest users of glass"*: Ford, introduction to *My Life and Work.*

51   *over the previous year:* Watts, *The People's Tycoon,* chapter 13.

51   *They filed a lawsuit:* Ibid.

51   *"believe that we should"*: Ibid.

51   *"Business is a service"*: Snow, *I Invented the Modern Age,* 257.

53   *to a Russian firm:* Gerald F. Davis, *Managed by the Markets: How Finance Reshaped America* (Oxford, UK: Oxford University Press, 2009), Preface.

53   *about 1,800 parts:* Sheffi, *The Magic Conveyor Belt,* 7.

53   *steering columns from Hungary:* J. B. Maverick, "Who Are Ford's Main Suppliers?" *Investopedia,* November 29, 2021.

53   *popular new pickup truck:* Michael Wayland, "Ford Quietly Begins Production of New 'C' Pickup in Mexico," CNBC, March 4, 2021.

53   *Nearly half the parts:* Data from US Department of Transportation, National Highway Traffic Safety Administration, "Part 583 American Automobile Labeling Act Reports," https://www.nhtsa.gov/part-583-american-automobile -labeling-act-reports.

53   *the computer chips:* Kyle Cheromcha, "Ford CEO Jim Farley Isn't Trying to Reinvent the Wheel—Just Everything Else Around It," *The Drive,* September 1, 2021, https://www.thedrive.com/news/42213/ford-ceo-jim-farley-isnt-trying -to-reinvent-the-wheel-just-everything-else-around-it.

54   *eight plants in North America:* Phoebe Wall Howard, "Ford Cuts Production at 8 Factories, Amid Semiconductor Chip Shortage, Alerts UAW," *Detroit Free Press,* June 30, 2021.

54   *"exactly the kind of thing"*: Interview with Matt Anderson, February 7, 2022.

54   *totaling $7.9 billion:* Analysis of publicly traded data by William Lazonick, an economist at the University of Massachusetts Lowell.

55   *"a Herculean task"*: Interview with Hau Thai-Tang, May 20, 2022.

CHAPTER 3: **"NO WASTE MORE TERRIBLE THAN OVERPRODUCTION"**

58   *"the fundamental ideas"*: James P. Womack, Daniel T. Jones, and Daniel Roos, *The Machine That Changed the World* (New York: Free Press, 1990), prologue.

59   *"The Toyota production system"*: Taiichi Ohno, *Toyota Production System: Beyond Large-Scale Production* (Boca Raton, FL: CRC Press, 1988), preface to the English edition; originally published in Japanese, *Toyota seisan hoshiki* (Tokyo: Diamond, 1978).

59   *"As in everything else"*: Ibid., chapter 5.

60   *in 1912 in Dalian:* Koichi Shimokawa and Takahiro Fujimoto, eds., *The Birth of*

Lean: Conversations with Taiichi Ohno, Eiji Toyoda and Other Figures Who Shaped Toyota Management (Cambridge, MA: Lean Enterprise Institute, 2009), chapter 1.

60   *thousand trucks per month:* Ibid.

61   *toward bankruptcy:* Ibid.

61   *American labor laws:* Womack et al., *The Machine That Changed the World,* chapter 3.

61   *"needed to find a way":* Shimokawa and Fujimoto, *The Birth of Lean,* chapter 1.

61   *the United States in 1910:* Ohno, *Toyota Production System,* chapter 4.

61   *"just in time":* Ibid.

62   *Ford's Rouge plant:* Shimokawa and Fujimoto, *The Birth of Lean,* chapter 5.

63   *"There is no waste":* Ohno, *Toyota Production System,* chapter 1.

63   *By the early 1980s:* Womack et al., *The Machine That Changed the World,* chapter 3.

64   *the largest manufacturers:* Marc Levinson, *The Box: How the Shipping Container Made the World Smaller and the World Economy Bigger* (Princeton, NJ: Princeton University Press, 2016), 356–57.

64   *"grew rice for subsistence":* Ohno, *Toyota Production System,* chapter 1.

CHAPTER 4: "THE LEAN TALIBAN"

66   *counseled the Saudi monarchy:* Katie Benner, Mark Mazzetti, Ben Hubbard, and Mike Isaac, "Saudi's Image Makers: A Troll Army and a Twitter Insider," *New York Times,* October 21, 2018.

66   *linked to Vladimir Putin:* Dan De Luce and Yasmine Salam, "McKinsey & Co. Worked with Russian Weapons Maker Even as It Advised Pentagon," NBC News, May 21, 2022, https://www.nbcnews.com/politics/national-security /consulting-firm-mckinsey-co-advised-state-owned-russian-defense-firm-r -rcna29618.

66   *to the Communist Party:* Walt Bogdanich and Michael Forsythe, "How McKinsey Has Helped Raise the Stature of Authoritarian Governments," *New York Times,* December 16, 2018.

66   *the Central Intelligence Agency:* Walt Bogdanich and Michael Forsythe, *When McKinsey Comes to Town: The Hidden Influence of the World's Most Powerful Consulting Firm* (New York: Doubleday, 2022), 3.

66   *"remake the world":* Ibid., 261.

67   *"the notice period":* Knut Alicke and Martin Lösch, "Lean and Mean: How Does Your Supply Chain Shape Up?" McKinsey & Company, 2010, 7.

67   *"super-flex temporary workforce":* Ibid.

67   *undertaken a survey:* Louis Hyman, *Temp: The Real Story of What Happened to Your Salary, Benefits, and Job Security* (New York: Penguin, 2019), chapter 10.

68   *"of continuous improvement":* Ibid.

68   *"reduce the uncertainty":* Knut Alicke, Daniel Rexhausen, and Andreas Seyfert, "Supply Chain 4.0 in Consumer Goods," McKinsey & Company, April 6, 2017, https://www.mckinsey.com/industries/consumer-packaged-goods/our -insights/supply-chain-4-0-in-consumer-goods.

69   *"supply chain organization":* Knut Alicke, Elena Dumitrescu, Christoph Lennartz, and Markus Leopoldseder, "New Organizational Wires for Digital Supply

Chains," McKinsey & Company, April 5, 2018, https://www.mckinsey.com /business-functions/operations/our-insights/new-organizational-wires-for -digital-supply-chains.

69 *"fundamentally evil":* Adam Lashinsky, "Tim Cook: The Genius Behind Steve," *Fortune,* August 24, 2011.

69 *"want to manage it":* Ibid.

70 *"keep reducing inventory":* Interview with ManMohan S. Sodhi, April 20, 2021.

70 *Between 1981 and 2000:* Hong Chen, Murray Z. Frank, and Owen Q. Wu, "What Actually Happened to the Inventories of American Companies Between 1981 and 2000?" *Management Science* 51, no. 7 (July 1, 2005), https://pubsonline .informs.org/doi/10.1287/mnsc.1050.0368.

70 *roughly $1.2 trillion:* Marc Levinson, *The Box: How the Shipping Container Made the World Smaller and the World Economy Bigger* (Princeton, NJ: Princeton University Press, 2016), 359.

70 *sometimes tied directly:* Bogdanich and Forsythe, *When McKinsey Comes to Town,* 7–8.

74 *more than $6 trillion:* Sirio Aramonte, "Mind the Buybacks, Beware of the Lever-age," Bank for International Settlements, *BIS Quarterly Review,* September 14, 2020, https://www.bis.org/publ/qtrpdf/r_qt2009d.htm.

74 *Japan, Britain, France:* Ibid.

74 *rattled Taiwan:* Peter S. Goodman, "Tech Stocks Swoon in Wake of Taiwan Quake," *Washington Post,* September 22, 1999.

74 *locate popular products:* Rachel Beck, "Parents Scrambling for Popular Toys," As-sociated Press, December 22, 1999.

75 *"a global production system":* Barry C. Lynn, *End of the Line: The Rise and Coming Fall of the Global Corporation* (New York: Doubleday, 2005), 3.

76 *26 percent stake in J.C. Penney:* Rachel Dodes and Joann S. Lublin, "J.C. Penney, Facing Activist Efforts, Adopts 'Poison Pill,'" *Wall Street Journal,* October 18, 2010.

76 *"Steve Jobs of the retail industry":* Michael J. de la Merced, "Ackman Scores Big Win with Penney's Latest Hire," *New York Times,* June 14, 2011.

76 *free up $500 million:* Transcript of J.C. Penney Inc. Earnings Conference Call, Q2 2012, August 10, 2012, CQ Transcriptions.

76 *a year and a half:* Emily Glazer, Joann S. Lublin, and Dana Mattioli, "Penney Backfires on Ackman," *Wall Street Journal,* April 9, 2013.

76 *lost their lives:* Vasco M. Carvalho, Makoto Nirei, Yukiko U. Saito, and Alireza Tahbaz-Salehi, "Supply Chain Disruptions: Evidence from the Great East Japan Earthquake," *Quarterly Journal of Economics* (2021): 1255–321.

77 *advanced silicon wafers:* Don Lee and David Pierson, "Disaster in Japan Exposes Supply Chain Flaw," *Los Angeles Times,* April 6, 2011.

77 *shortage of electronic sensors:* Ibid.

77 *Peugeot-Citroën:* Ibid.

77 *"There should have been fail-safe measures":* Ibid.

77 *"We'll do a retrospective":* Steve Lohr, "Stress Test for the Global Supply Chain," *New York Times,* March 20, 2011.

78 *"a pandemic will strike":* Ian Goldin and Mike Mariathasan, *The Butterfly Defect:*

How Globalization Creates Systemic Risks, and What to Do About It (Princeton, NJ: Princeton University Press, 2014), chapter 6.

78 *Between 2009 and 2018:* William Lazonick, Mustafa Erdem Sakinç, and Matt Hopkins, "Why Stock Buybacks Are Dangerous for the Economy," *Harvard Business Review* (January 7, 2020), https://hbr.org/2020/01/why-stock-buybacks-are-dangerous-for-the-economy.

78 *reduced their inventories:* "Manufacturers: Inventories to Sales Ratio," data, as captured by United States Census Bureau manufacturing surveys, tabulated by St. Louis Fed, https://fred.stlouisfed.org/series/MNFCTRIRSA.

80 *"getting starved here":* Telephone interview with Farrell, October 22, 2021.

81 *"money controls everything":* Telephone interview with Joseph Norwood, November 9, 2021.

CHAPTER 5: **"EVERYBODY WANTS EVERYTHING."**

85 *more than eighty thousand:* Donald G. McNeil Jr., "The U.S. Now Leads the World in Confirmed Coronavirus Cases," *New York Times,* March 26, 2020.

85 *since the Great Depression:* Heather Long and Andrew Van Dam, "U.S. Unemployment Rate Soars to 14.7 Percent, the Worst Since the Depression Era," *Washington Post,* May 8, 2020.

85 *In Europe, joblessness rose:* Peter S. Goodman, Patricia Cohen, and Rachel Chaundler, "European Workers Draw Paychecks. American Workers Scrounge for Food," *New York Times,* July 4, 2020, A1.

86 *an astonishing 32.9 percent:* Ben Casselman, "A Collapse That Wiped Out 5 Years of Growth, with No Bounce in Sight," *New York Times,* July 31, 2020, A1.

86 *"This is a crisis like no other":* Gita Gopinath, "The Great Lockdown: Worst Economic Downturn Since the Great Depression," *IMFBlog,* April 14, 2020.

86 *Major apparel brands:* "FACTBOX – Fashion brands cut orders with Asian garment makers," Reuters, May 18, 2020.

86 *broadly trimmed orders:* "U.S. Core Capital Goods Orders Point to Worsening Business Investment Downturn," Reuters, March 25, 2020.

86 *the newest iPhones:* Tripp Mickle, "Would You Buy an iPhone Now? Coronavirus Tests Demand for Apple's Flagship Product," *Wall Street Journal,* March 26, 2020.

86 *seventy scheduled runs:* "919 blank sailings on Transpacific and Asia-Europe," Sea-Intelligence, May 4, 2021, https://www.sea-intelligence.com/press-room/65-919-blank-sailings-on-transpacific-and-asia-europe.

86 *126 scheduled sailings:* US International Trade Commission, "The Impact of the Covid-19 Pandemic on Freight Transportation Services and U.S. Merchandise Imports," https://www.usitc.gov/research_and_analysis/tradeshifts/2020/special_topic.html.

87 *waters of Southeast Asia:* Costas Paris, "Ocean Carriers Idle Container Ships in Droves on Falling Trade Demand," *Wall Street Journal,* April 8, 2020.

87 *recycling operations:* Costas Paris, "Cargo Vessels and Cruise Ships Line Up for Scrapping," *Wall Street Journal,* November 10, 2020.

87 *between 15 and 30 percent:* US International Trade Commission, "The Impact of the Covid-19 Pandemic," fn. 9.

87   *global freight-carrying capacity:* Data furnished by Everstream Analytics.

87   *"It's very quiet":* Margot Roosevelt, "Workers Suffer as Traffic at Ports Drops," *Los Angeles Times,* March 7, 2020, A1.

88   *more than one-fifth:* US Department of Commerce data tabulated by Deloitte Services LP, as cited in Akrur Barua, "A Spring in Consumers' Steps," *Deloitte Insights,* June 2021, https://www2.deloitte.com/us/en/insights/economy/us -consumer-spending-after-covid.html.

88   *recreational goods and vehicles:* Ibid.

88   *"Everybody wants everything":* Interview with Akhil Nair, February 16, 2021.

89   *launch new businesses:* Dali L. Yang, *Remaking the Chinese Leviathan: Market Transition and the Politics of Governance in China,* (Stanford, CA: Stanford University Press, 2004), 155–58.

89   *$2 billion a year:* Ningbo Municipal Bureau of Statistics data, tabulated by CEIC, https://www.ceicdata.com/en/china/foreign-direct-investment-capital -utilized-prefecture-level-city/cn-fdi-utilized-zhejiang-ningbo.

89   *third largest container port:* "The Top 50 Container Ports," World Shipping Council, https://www.worldshipping.org/top-50-ports.

CHAPTER 6: "AN ENTIRE NEW WAY OF HANDLING FREIGHT"

91   *more than 1,700 vehicles:* Anthony J. Mayo and Nitin Nohria, "The Truck Driver Who Reinvented Shipping," Harvard Business School, Working Knowledge, October 3, 2005, https://hbswk.hbs.edu/item/the-truck-driver-who-reinvented -shipping.

92   *hundreds of cargo ships:* Marc Levinson, *The Box: How the Shipping Container Made the World Smaller and the World Economy Bigger* (Princeton, NJ: Princeton University Press, 2016), 57.

93   *over the centuries:* Daniel M. Bernhofen, Zouheir El-Sahli, and Richard Kneller, "Estimating the Effects of the Container Revolution on World Trade," Center for Economic Studies & Ifo Institute, Working Paper No. 4136, Munich, Germany, February 2013.

93   *riotous assortment of cargo:* Levinson, *The Box,* 22.

94   *Between 1870 and 1910:* Kevin H. O'Rourke and Jeffrey G. Williamson, "When Did Globalization Begin?" National Bureau of Economic Research, Working Paper No. 7632, April 2000, Table 1.

94   *the construction of telegraph wires:* Helena Vieira, "The Trade Impact of the Transatlantic Telegraph," *LSE Business Review* (blog), London School of Economics and Political Science, March 20, 2018, https://blogs.lse.ac.uk/business review/2018/03/20/the-trade-impact-of-the-transatlantc-telegraph/.

94   *end of the twentieth century:* Levinson, *The Box,* 360–61.

95   *"into a factory job":* Ibid., 169.

95   *"much nostalgia":* Marian Betancourt, *Heroes of New York Harbor: Tales From the City's Port* (Guilford, CT: GlobePequot, 2017), chapter 10.

95   *the Motor Carrier Act in 1930:* Marco Poisler and Edward D. Greenberg, "History of Trucking Regulation: 1935 to 1980," Transportation Lawyers Association.

96 *on the other hand:* Levinson, *The Box,* 57.

96 *undercutting them on price:* Ibid., 58.

96 *Ballantine Beer:* Ibid., 63.

96 *a manufacturer in Spokane:* Ibid., 66.

96 *The docks of New York:* Ibid., 103–4.

97 *had abundant room:* Ibid., 58.

97 *local officials gathered:* "Tankers to Carry 2-Way Pay Loads," *New York Times,* April 27, 1956, 39.

97 *"sink that sonofabitch":* CQ Researcher, ed., *Issues in Terrorism and Homeland Security: Selections from CQ Researcher,* 2nd ed. (Thousand Oaks, CA: Sage Publications, 2011), 326.

97 *costs of moving freight:* Levinson, 27.

97 *nearly $6 a ton:* Betancourt, *Heroes of New York Harbor,* chapter 10.

98 *476 containers:* Levinson, *The Box,* 94.

98 *opened a terminal:* Betancourt, *Heroes of New York Harbor,* chapter 10.

98 *400,000 additional troops: The Pentagon Papers: The Secret History of the Vietnam War* (New York: Racehorse, 2017), chapter 8.

98 *lone deepwater port:* Levinson, *The Box,* 232.

99 *notoriously shallow:* Ibid.

99 *McLean persuaded:* Ibid., 239.

99 *a $70 million contract:* Ibid., 243.

99 *The Japanese government was:* Ibid., 251–52.

99 *running six ships:* Ibid., 253.

100 *Between 1966 and 1983:* Bernhofen, El-Sahli, and Kneller, "Estimating the Effects," 12.

100 *nearly ninefold:* Ibid., 19.

100 *exceeding $100 million:* Hercules E. Haralambides, "Gigantism in Container Shipping, Ports and Global Logistics: A Time Lapse into the Future," *Maritime Economics & Logistics* 21 (January 2019): 1–60, https://doi.org/10.1057/s41278-018-00116-0.

100 *the consultants at McKinsey:* Levinson, *The Box,* 273.

100 *recovered from the devastation:* J. G. De Gijt, J. M. Van Kleef, P. Taneja, and Han Ligteringen, "Development of Container Handling in the Port of Rotterdam," ResearchGate, January 2010, https://www.researchgate.net/figure/Historical-development-of-Port-of-Rotterdam-Source-PoR_fig5_311981594.

100 *middle of the 1980s:* Jitendra Bhonsle, "Evolution and Upsizing of Container Vessels," *Marine Insight,* February 11, 2022.

101 *more than 23,000 containers:* "Top 10: The Largest Container Ships in the World," *Container News,* December 6, 2021, https://container-news.com/top-10-the-largest-container-ships-in-the-world/.

101 *fourth largest container carrier:* "2022 Top 50 Ocean Carriers," *American Journal of Transportation* 740 (April 18–May 15, 2022): 8.

101 *80 percent of the cranes:* Blanchette, "Hidden Harbors."

101 *ten busiest container ports:* "The Top 50 Container Ports," World Shipping Council, https://www.worldshipping.org/top-50-ports.

101 *roughly 50 million:* Data courtesy of Drewry.

CHAPTER 7: **"CARRIERS ARE ROBBING SHIPPERS."**

104 *"a bunch of containers"*: Peter S. Goodman, Alexandra Stevenson, Niraj Chokshi, and Michael Corkery, "'I've Never Seen Anything Like This': Chaos Strikes Global Shipping," *New York Times*, March 7, 2021, A1.

105 *"We have a dramatic situation"*: Interview with John Whelan, February 15, 2021.

105 *in Sub-Saharan Africa:* "Covid-19 and Maritime Transport: Impact and Responses," United Nations Conference on Trade and Development, Transport and Trade Facilitation, Series No. 15, 2021, 15.

106 *ordinary shipping containers:* Greg Miller, "How Three Chinese Companies Cornered Global Container Production," *American Shipper,* May 24, 2021, https://www.freightwaves.com/news/how-three-chinese-companies-cornered-global-container-production.

106 *China Container Industry Association:* David Dayen, "Rollups: A Chinese Corner in Chassis and Containers," *American Prospect,* April 6, 2022.

106 *Dong Fang International Containers:* "Company Profiles," Dong Fang International Containers (website), accessed September 8, 2022, https://www.dfichk.com/company-profiles.html.

106 *"behaving differently"*: Greg Miller, "Chinese Factories Won't Build Enough Boxes to Save US Shippers," *American Shipper,* February 17, 2021, https://www.freightwaves.com/news/chinese-factories-wont-build-enough-containers-to-save-us-shippers.

107 *"to try to support"*: Miller, "How Three Chinese Companies."

107 *"anything like this"*: Goodman et al., "'I've Never Seen Anything Like This.'"

107 *Between September and November:* Data compiled by Sea-Intelligence at author's request.

107 *a fresh outbreak:* Keith Bradsher, "Covid Crackdown at Chinese Port Stalls Global Trade," *New York Times,* June 22, 2021.

108 *Shipping Act of 1916:* Clarence G. Morse, "A Study of American Merchant Marine Legislation," *Law and Contemporary Problems* 25, no. 1 (Winter 1960): 57–58.

108 *American antitrust law:* Matt Stoller, "Too Big to Sail: How a Legal Revolution Clogged Our Ports," *BIG,* November 13, 2021, https://mattstoller.substack.com/p/too-big-to-sail-how-a-legal-revolution.

108 *"disapprove, cancel, or modify"*: The Shipping Act of 1916, Section 15, as cited in Susan Steinholtz Sennett, "Pre-Implementation Review Under Section 15 of the Shipping Act of 1916," *Loyola University Chicago Law Journal* 9, no. 1 (Fall 1977): 248.

109 *limited its authority:* "Competition Issues in Liner Shipping, United States," Organization for Economic Cooperation and Development, Competition Committee, Directorate for Financial and Enterprise Affairs, Working Party No. 2 on Competition and Regulation, June 19, 2015, 2.

109 *their prices and routes:* Ibid.

110 *"completely secret contracts"*: Statement of Harold J. Creel Jr. before the US Senate Commerce Committee, Subcommittee on Surface Transportation and Merchant Marine, March 20, 1997, Federal Document Clearinghouse.

110 *controlled 80 percent:* "The Impact of Alliances in Container Shipping," Inter-

national Transport Forum, Paris, 2018, Figure 1, 14, https://www.itf-oecd.org /sites/default/files/docs/impact-alliances-container-shipping.pdf.

111 *spurred domestic shipbuilding:* Marc Levinson, *Outside the Box: How Globalization Changed from Moving Stuff to Spreading Ideas* (Princeton, NJ: Princeton University Press, 2020), 206.

112 *of seaborne trade:* Vivian Yee and James Glanz, "How One of the World's Biggest Ships Jammed the Suez Canal," *New York Times,* July 18, 2021, A10.

112 *the International Transport Forum:* "The Impact of Mega-Ships," International Transport Forum, 2015, 9, https://www.itf-oecd.org/sites/default/files/docs /15cspa_mega-ships.pdf.

112 *smaller number of ports:* Ibid., 34–38.

112 *on a regular flow:* Interview with Marc Levinson on *Odd Lots* podcast, March 31, 2021.

113 *"hasn't really worried":* Telephone interview with Timothy Boyle, August 24, 2021.

114 *$300 billion in profits:* Greg Miller, "Despite Rising Risks, Shipping Lines on Track for Another Record Year," *American Shipper,* April 26, 2022.

114 *"manipulating the market":* Interview with Jason Delves, March 25, 2022.

114 *move 1,040 containers:* Data analysis furnished by F9 Brands.

116 *the International Monetary Fund estimated:* Yan Carriere-Swallow, Pragyan Deb, Davide Furceri, Daniel Jimenez, and Jonathan David Ostry, "Shipping Costs and Inflation," International Monetary Fund, Working Paper No. 2022/061, March 25, 2022, https://www.imf.org/en/Publications/WP/Issues /2022/03/25/Shipping-Costs-and-Inflation-515144.

116 *"We are finding it impossible":* Interview with David Reich, April 26, 2022.

117 *"massive competitive advantage":* Greg Miller, "Beware 'Nasty Side Effects' If Government Targets Ocean Carriers," *American Shipper,* August 5, 2021, https:// www.freightwaves.com/news/beware-nasty-side-effects-if-government-targets -ocean-carriers.

120 *shipping containers worldwide:* Greg Holt, "MSC Tops Maersk to Become World's Largest Container Shipping Line: Alphaliner," S&P Global, Commodity Insights, January 5, 2022, https://www.spglobal.com/commodityinsights/en/market -insights/latest-news/shipping/010522-msc-tops-maersk-to-become-worlds -largest-container-shipping-line-alphaliner.

120 *seven hundred vessels:* "How We Keep Global Trade Moving," MSC, https://www .msc.com/en/about-us.

CHAPTER 8: **"THE LAND OF THE FORGOTTEN"**

126 *leaving the ports:* Data compiled by Sea-Intelligence at author's request.

127 *increasingly bypassing Oakland:* Ibid.

127 *1.1 billion pounds of almonds:* Almond Alliance of California data.

127 *"actively looking to investigate cases":* Interview with Dan Maffei, April 5, 2022.

128 *"don't have to compete":* The White House, "Remarks of President Joe Biden— State of the Union Address as Prepared for Delivery," March 1, 2022, https:// www.whitehouse.gov/briefing-room/speeches-remarks/2022/03/01/remarks -of-president-joe-biden-state-of-the-union-address-as-delivered/.

130   *a gallon of water:* Tom Philpott, "Invasion of the Hedge Fund Almonds,"
      Mother Jones (January 12, 2015), https://www.motherjones.com/environment
      /2015/01/california-drought-almonds-water-use/.

130   *businesses in Los Angeles:* Julia Lurie, "California's Almonds Suck as Much Water
      Annually as Los Angeles Uses in Three Years," *Mother Jones* (January 12, 2015),
      https://www.motherjones.com/environment/2015/01/almonds-nuts-crazy
      -stats-charts/.

130   *its largest industry was agriculture:* Sona P., "The 7 Biggest Industries in Cali-
      fornia," California.com, June 23, 2021, https://www.california.com/biggest
      -industries-california/.

130   *a fifth of that total:* "California Agricultural Exports, 2020–2021," California Ag-
      ricultural Statistics Review, California Department of Food & Agriculture, 1.

CHAPTER 9: **"I THINK I'VE HEARD OF THEM."**

139   *"taking advantage":* The White House, "President Biden Signs into Law S. 3580,
      the Ocean Shipping Reform Act of 2022," YouTube video, June 16, 2022, https://
      www.youtube.com/watch?v=juG_UDFQqaE.

140   *$14 trillion worth of products:* "Shipping and World Trade: Driving Prosperity," In-
      ternational Chamber of Shipping, https://www.ics-shipping.org/shipping-fact
      /shipping-and-world-trade-driving-prosperity/.

141   *"hostage to the industry":* Interview with Peter Friedmann, July 28, 2022.

141   *shipping interests in China:* Michael Forsythe and Eric Lipton, "For the Chao Fam-
      ily, Deep Ties to the World's Two Largest Economies," *New York Times,* June 2,
      2019.

142   *"I'm just extraordinarily ordinary":* Mark Weiner, "Dan Maffei; A Brainy Kid Comes
      Home and Seeks a Return to D.C.," *Syracuse Post-Standard,* September 9, 2008, A1.

143   *"it was going to happen":* Michael Angell, "Ocean Carriers Pledge Efforts to Hit
      OSRA Export Mandates," *Journal of Commerce,* June 21, 2022, https://www
      .joc.com/maritime-news/container-lines/mediterranean-shipping-co/ocean
      -carriers-pledge-efforts-hit-osra-export-mandates_20220621.html.

144   *"disturbingly high by historical measures":* "Effects of the Covid-19 Pandemic on
      the U.S. International Ocean Supply Chain," Federal Maritime Commission,
      Fact Finding Investigation 29, Final Report, May 31, 2022, 6.

150   *"in any renewal discussion":* Internal email from Juergen Pump to Kevin Li,
      April 29, 2021, attached as exhibit to "Complainants Memorandum of Law in
      Opposition to Respondents' Partial Motion to Dismiss And/Or For Summary
      Judgement," in OJ Commerce, LLC, Complainant, v. Hamburg Südamerikan-
      ische Dampfschifffahrts-Gesellschaft A/S & CO KG and Hamburg Süd North
      America, Inc., Respondents, Federal Maritime Commission, Docket No. 21-11,
      August 10, 2022, 11.

150   *"The clear intent of Congress":* Telephone interview with Dan Maffei, November 4,
      2022.

151   *ignored her directives:* "Order on Respondents' Motion to Partially Dismiss and
      for a Protective Order and Complainant's Motion for Expedited Relief," OJ Com-
      merce v. Hamburg Südamerikanische, August 31, 2022.

CHAPTER 10: **"EVERYTHING IS OUT OF WHACK."**

153 *increase by 16 percent:* Brandon Richardson, "Ports of Long Beach, Los Angeles Shatter Annual Cargo Volume Record," *Long Beach Business Journal,* January 20, 2022, https://lbbusinessjournal.com/ports/port-of-long-beach-shatters-annual-cargo-volume-record.

153 *exceeded fifty vessels:* Data furnished by Adil Ashiq at AIS Marine Intelligence, provided by MarineTraffic.

153 *For the first six days:* Ibid.

153 *Dead ahead of the* Emden: Ibid.

154 *waiting nearly two weeks:* Jenny Leonard, "U.S. Consumer Watchdog Probes Supply Logjams Fueling Inflation," *Bloomberg,* November 30, 2021.

154 *two-thirds of their capacity:* Costas Paris and Jennifer Smith, "Cargo Piles Up as California Ports Jostle Over How to Resolve Delays," *Wall Street Journal,* September 26, 2021.

154 *"has the potential":* Chris Megerian and Don Lee, "Biden Takes on Supply Chain Logjam," *Los Angeles Times,* October 14, 2021, A1.

155 *1 percent and below:* Data furnished by Prologis.

156 *"have been impacted":* Emailed statement from Tom Boyd, March 23, 2022.

156 *$62 billion in revenue:* Maersk press release, February 9, 2022.

156 *"the best quarter":* Greg Miller, "Shipping Giant Maersk Continues Buying Spree After Best Quarter Ever," *FreightWaves,* November 2, 2021, https://www.freightwaves.com/news/shipping-giant-maersk-continues-buying-spree-after-best-quarter-ever.

157 *"Every terminal should be working":* Interview with Jesse Lopez in Long Beach, California, March 7, 2022.

157 *"there's much incentive":* Interview with Jaime Hipsher in Long Beach, California, March 7, 2022.

157 *revenue reaching $21.7 billion:* "A.P. Møller—Maersk A/S (AMKBY) CEO Soren Skou on Q2 2022 Results—Earnings Call Transcript," Seeking Alpha, August 3, 2022, https://seekingalpha.com/article/4529361-p-moller-maersk-s-amkby-ceo-soren-skou-on-q2-2022-results-earnings-call-transcript.

157 *"We will provide the capacity":* Ibid.

158 *"has been crooked":* Interview with Gene Seroka in Los Angeles, March 8, 2022.

158 *fifty vessels were stuck:* Aaron Clark and Kevin Varley, "Cyclone Closes One of World's Busiest Ports, Creating Ship Traffic Jam," *Bloomberg,* October 12, 2021.

158 *floating off the port of Ningbo:* Greg Miller, "California Congestion Nears New High, East Coast Gridlock Worsens," *FreightWaves,* August 16, 2021.

160 *"to get their freight":* Interview with Griff Lynch in Savannah, Georgia, September 29, 2021.

CHAPTER 11: **"CRAZY AND DANGEROUS"**

163 *"it was known as shape-up":* Marc Levinson, *The Box: How the Shipping Container Made the World Smaller and the World Economy Bigger* (Princeton: Princeton University Press, 2016), 29.

163 *often borrowing money:* Ibid.

164 *in Britain and Australia:* Ibid., 29–30.

167 *the work stoppage:* Ibid., 143.

167 *along the West Coast:* Ibid., 29.

168 *"like we're nobodies":* Interview with Anthony Chilton, Long Beach, California, March 10, 2022.

168 *"rudest people in the world":* Interview with Marshawn Jackson in Los Angeles, September 13, 2022

169 *at least two dozen members:* Data furnished by ILWU.

170 *represented by a union:* US Bureau of Labor Statistics, "Union Members Summary," press release, January 19, 2023, https://www.bls.gov/news.release /union2.nr0.htm.

170 *20 percent in 1983:* Ibid.

170 *28 percent in 1954:* Gerald Mayer, "Union Membership Trends in the United States," Congressional Research Service, Domestic Social Policy Division, August 31, 2004.

170 *10 and 30 percent more:* Ibid.

170 *angering the rank and file:* Qianqian Huang, Feng Jiang, Erik Lie, and Tingting Que, "The Effect of Labor Unions on CEO Compensation," *Journal of Financial and Quantitative Analysis* 52, no. 2 (April 2017): 553–82.

170 *extravagant raises and bonuses:* Josh Bivens and Jori Kandra, "CEO Pay Has Skyrocketed 1,460% Since 1978," Economic Policy Institute, October 4, 2022, https://www.epi.org/publication/ceo-pay-in-2021/.

170 *four hundred times as much:* Ibid.

171 *three-fourths of those families:* Tracy A. Loveless, "About a Third of Families Who Received Supplemental Nutrition Assistance Program Benefits Had Two or More People Working," United States Census Bureau, July 21, 2020.

171 *some form of government benefits:* Nir Kaissar and Timothy L. O'Brien, "Who Helps Pay Amazon's Low Wage Workers? You Do," *Bloomberg*, March 18, 2021.

172 *"reaching an agreement":* John Drake, "How Ongoing Labor Negotiations Are Impacting Inflation and Supply Chains," US Chamber of Commerce, September 7, 2022, https://www.uschamber.com/security/supply-chain/how-ongoing-labor -negotiations-are-impacting-inflation-and-supply-chains.

172 *"labor disputes at ports":* "Trade Association Letter to White House," Retail Industry Leaders Association, March 1, 2022, https://www.rila.org/focus-areas /supply-chain/trade-association-letter-port-negotiations.

173 *the costs of ocean shipping:* Levinson, *The Box,* 27.

173 *a controversial deal:* Andrea Hsu, "Retired Labor Leader Says His Former Union Must Think Outside the Box to Save Jobs," NPR, September 11, 2022 https:// www.npr.org/2022/09/11/1121373064/shipping-union-dockworker-ilwu -supply-chain-automation.

174 *"it was physically taxing":* Interview with John Arkenbout in Rotterdam, July 12, 2016.

CHAPTER 12: "IS IT WORTH EVEN GETTING UP IN THE MORNING?"

180 *"How do you convince truckers"*: Ryan Johnson, "I'm a Twenty Year Truck Driver, I Will Tell You Why America's 'Shipping Crisis' Will Not End," *Medium*, October 27, 2021, https://medium.com/@ryan79z28/im-a-twenty-year-truck-driver-i-will -tell-you-why-america-s-shipping-crisis-will-not-end-bbe0ebac6a91.

180 *dray operators were quitting*: Ari Ashe, "US Drayage Drivers Quitting as Rail Ramp Congestion Crimps Pay," *Journal of Commerce*, May 19, 2021.

180 *at the port in Charleston*: David Wren, "Anchored Ship Logjam Increases off SC Coast as Charleston Port Sees Container Overload," *Post and Courier*, February 15, 2022.

180 *one-fourth of households*: Ruby Bolaria, Alexis Cooke, and Natalie Nava, "South Central Neighborhood Council Report," December 20, 2013, https://vnnc.org /wp-content/uploads/2014/05/The-State-of-South-Central-NC.pdf.

181 *"this country would stop"*: TheVideoWhisperer, "Trucking in America—Driving for Swift, Excerpts," 2010, YouTube video, https://www.youtube.com/watch?v =3t4i8kfftFc.

182 *"Many of the workers"*: Steve Viscelli, *The Big Rig: Trucking and the Decline of the American Dream* (Oakland: University of California Press, 2016), chapter 1.

189 *Of the fifty-eight thousand chassis*: Aurora Armendral, "Unloading Container Ships Faster in the US Is Pushing Supply Chain Woes onto Trucks," *Quartz*, October 18, 2021.

CHAPTER 13: "BUILDING RAILROADS FROM NOWHERE TO NOWHERE AT PUBLIC EXPENSE"

193 *the greenhouse gas emissions*: Alexander Laska, "Freight Rail's Role in a Net-Zero Economy," *Third Way*, June 7, 2021.

193 *as transporting it by train*: David Austin, "Pricing Freight Transport to Account for External Costs," Congressional Budget Office, Working Paper No. 2015-03, March 2015, 35.

194 *eight times greater*: Ibid., Summary.

195 *new methods of management*: Matt Stoller, *Goliath: The 100-Year War Between Monopoly Power and Democracy* (New York: Simon & Schuster, 2019), 6–7.

195 *inaugural run in 1830*: "Baltimore and Ohio Railroad," *Encyclopedia.com*, https:// www.encyclopedia.com/history/encyclopedias-almanacs-transcripts-and-maps /baltimore-and-ohio-railroad.

196 *cities in the North*: Michael Hiltzik, *Iron Empires: Robber Barons, Railroads, and the Making of Modern America* (New York: Mariner Books, 2020), chapter 1.

196 *to move in troops*: Sam Vong, "The Impact of the Transcontinental Railroad on Native Americans," *O Say Can You See?* (blog), National Museum of American History, June 3, 2019. https://americanhistory.si.edu/blog/TRR.

196 *Trains carried farming equipment*: *American Experience*, season 15, episode 7, "The Transcontinental Railroad," aired January 27, 2003, on PBS.

196 *"Time and space are annihilated"*: David Haward Bain, *Empire Express: Building the First Transcontinental Railroad* (New York: Penguin Books, 1999), 8–9.

196 *363-mile journey from Albany*: Hiltzik, *Iron Empires,* chapter 1.

197 *began recruiting Chinese*: Gordon H. Chang, Shelly Fisher Fishkin, and Hilton Obenzinger, introduction to *The Chinese and the Iron Road: Building the Transcontinental Railroad,* eds. Gordon H. Chang and Shelley Fisher Fishkin (Stanford, CA: Stanford University Press, 2019).

197 *more than twelve thousand*: Bain, *Empire Express,* 209.

197 *organized recruitment system*: Evelyn Hu-DeHart, "Chinese Labor Migrants to the Americas in the Nineteenth Century," in *The Chinese and the Iron Road.*

197 *nearly in half*: Chang, Fiskin, and Obenzinger, introduction to *The Chinese and the Iron Road.*

197 *record their names*: Barbara L. Voss, "Living Between Misery and Triumph," in *The Chinese and the Iron Road.*

197 *the riskiest tasks*: J. Ryan Kennedy, Sarah Heffner, Virginia Popper et al., "The Health and Well-Being of Chinese Railroad Workers," in *The Chinese and the Iron Road.*

197 *rest of the operations*: Voss, "Living Between Misery and Triumph."

198 *"are Chinese"*: Bain, *Empire Express,* 219–20.

198 *took to the streets*: Ibid., 666–67.

198 *"The explosion of railroad building"*: Hiltzik, *Iron Empires,* chapter 1.

199 *grants of land*: Maury Klein, "Financing the Transcontinental Railroad," Gilder Lehrman Institute of American History, https://ap.gilderlehrman.org/essays/financing-transcontinental-railroad.

199 *Union Pacific*: Hiltzik, *Iron Empires,* chapter 3.

199 *comprehensive sprinkling of shares*: Bain, *Empire Express,* 679–80.

199 *some $20 million*: Hiltzik, *Iron Empires,* chapter 4.

200 *"Building railroads from nowhere"*: Ibid., chapter 5.

200 *American railroads to bankruptcy*: Ibid., chapter 4.

200 *to eliminate competition*: Stoller, *Goliath,* 9–10.

200 *American business lexicon*: Ibid., 11.

201 *demand regulation of rates*: Ibid., 9.

201 *"an unskilled cipher"*: Robert V. Bruce, *1877: Year of Violence* (Indianapolis: Bobbs-Merrill Co., 1959), chapter 1.

201 *"A brakeman with both hands"*: Ibid., chapter 3.

202 *"The regular compensation of employees"*: Ibid.

202 *the layover*: Ibid.

202 *The Pennsylvania slashed*: State of Pennsylvania, General Assembly, *Report of the Committee Appointed to Investigate the Railroad Riots in July, 1877* (Harrisburg, PA: Lane S. Hart, State Printer, 1878), https://www.google.com/books/edition/Report_of_the_Committee_Appointed_to_Inv/jtE5AQAAIAAJ?hl=en.

202 *its earnings in dividends*: Bruce, *1877,* chapter 3.

203 *"a rifle diet"*: Stoller, *Goliath,* 11.

203 *train cars and locomotives*: Bruce, *1877,* chapter 8.

203 *"were all the necessities"*: Ibid.

203 *more than five thousand*: Ibid.

203   *two dozen people:* Record Group 10, records of the office of Pennsylvania Governor John F. Hartranft and photos of the scene, "Description of Railroad Riots— July 23, 1877," Pennsylvania Historical & Museum Commission, http://www.phmc.state.pa.us/portal/communities/documents/1865-1945/railroad-riots.html.

203   *orange glow on the horizon:* Bruce, *1877*, chapter 8.

203   *an "insurrection":* "Hayes' July 21 Proclamation: A Manifesto Against Domestic Violence," Railroads and the Making of Modern America, University of Nebraska, Lincoln, https://railroads.unl.edu/documents/view_document.php?keyword=hayes&id=rail.str.0016.

204   *"put down by force":* "Rutherford B. Hayes Diary Entry, August 5, 1877," Railroads and the Making of Modern America, University of Nebraska, Lincoln, https://railroads.unl.edu/documents/view_document.php?rends%5B%5D=diary&sort=title&page=1&order=desc&id=rail.str.0044.

206   *the Interstate Commerce Commission:* Stoller, *Goliath,* 9–10.

206   *a wave of consolidation:* Ibid.

206   *President Eisenhower embarked:* "Interstate Highway System," Dwight D. Eisenhower Presidential Library, Museum & Boyhood Home, https://www.eisenhowerlibrary.gov/research/online-documents/interstate-highway-system.

206   *end of World War II:* Matthew Jinoo Buck, "How America's Supply Chains Got Railroaded," *American Prospect,* February 4, 2022, https://prospect.org/economy/how-americas-supply-chains-got-railroaded/.

206   *even less profitable routes:* Ibid.

206   *the American train system:* Ibid.

207   *"needless and costly regulation":* B. Kelly Eakin, A. Thomas Bozzo, Mark E. Meitzen, and Philip E. Schoech, "Railroad Performance Under the Staggers Act," *Regulation* (Winter 2010–2011): 32.

207   *"It will benefit shippers":* Ibid.

207   *"will have fewer lines":* "Rail Deregulation Passes House," *New York Times,* September 10, 1980, A7.

207   *the rise of retail giants:* Buck, "How America's Supply Chains."

207   *the first three decades:* Eakin et al., "Railroad Performance," 34.

208   *increase in "traffic density":* Ibid.

208   *every railroad merger:* Buck, "How America's Supply Chains."

208   *to a mere seven:* Marvin Prater, Adam Sparger, Daniel O'Neill Jr., "Railroad Concentration, Market Share and Rates," United States Department of Agriculture, Agricultural Marketing Service, February 2014.

208   *from 53 percent to 86 percent:* Ibid.

208   *the freight stations:* Martha Moore, "U.S. Freight Customers Increasingly Taxed by Higher Rail Rates," *Regulatory Review,* June 24, 2019, https://www.theregreview.org/2019/06/24/moore-us-freight-customers-taxed-higher-rail-rates/.

208   *first fourteen years:* Eakin et al., "Railroad Performance."

208   *by more than half:* Bureau of Transportation Statistics data, cited in Buck, "How America's Supply Chains."

210   *nearly tripled:* Moore, "U.S. Freight Customers."

CHAPTER 14: "THE ALMIGHTY OPERATING RATIO"

209   *by 29 percent:* Testimony of Martin J. Oberman, chairman, Surface Transportation Board, in hearing before US House of Representatives, Committee on Transportation and Infrastructure, Subcommittee on Railroads, Pipelines and Hazardous Materials, May 12, 2022.

210   *"what is driving all this":* Matthew DeLay, "'It Is Getting Worse. People Are Leaving,'" *Railway Age,* April 25, 2022, https://www.railwayage.com/regulatory/it-is-getting-worse-people-are-leaving/.

211   *"streamline operations":* "Chicago Intermodal Simplification and Service Update from Kenny Rocker, EVP, Marketing & Sales," announcement no. CN2019-28, Union Pacific, May 2, 2019, https://www.up.com/customers/announcements/customernews/generalannouncements/CN2019-28.html.

211   *by one-fifth:* Union Pacific Corp. Form 10-K, 2020, 32.

211   *$2.6 billion in dividends:* Ibid., 3–4.

211   *"inland intermodal terminals":* Quoted in Ari Ashe, "UP Suspending USWC-Chicago Hub Services to Clear Global IV Boxes," *Journal of Commerce* (July 15, 2021).

211   *coming in from Chicago:* Bill Mongelluzzo, "UP Suspends Receipt of Westbound Boxes at LA-LB Facility," *Journal of Commerce* (October 25, 2021).

212   *"How many days or weeks":* Letter from Michael Paul Lindsey II to the Surface Transportation Board, July 18, 2022.

212   *The culprit was Precision:* Ibid.

213   *adopted Precision in October 2018:* Jeff Stagl, "It's Taking a Team Effort for Union Pacific to Roll Out Its Version of PSR," *Progressive Railroading,* March 2019, https://www.progressiverailroading.com/union_pacific/article/Its-taking-a-team-effort-for-Union-Pacific-to-roll-out-its-version-of-PSR--56940.

213   *more than $14 million:* Data compiled by Salary.com from Union Pacific proxy statements.

213   *$2.7 million in compensation:* Ibid.

214   *"for chemical shippers":* Testimony from Chris Jahn, president of the American Chemistry Council, before the Surface Transportation Board, April 26, 2022, https://www.americanchemistry.com/content/download/10951/file/ACC-Testimony-to-STB-on-Urgent-Issues-in-Freight-Rail-Service-042622.pdf.

214   *The fertilizer industry:* Letter from Corey Rosenbush, president and CEO of The Fertilizer Institute, to Surface Transportation Board, June 2, 2021, https://www.stb.gov/wp-content/uploads/Fertilizer-Institute-to-Board-relating-to-CSX_20210602.pdf.

215   *"There isn't any consistency":* Interview with David Heide in Kansas City, November 17, 2021.

215   *"astounding $183 billion":* Speech by Martin J. Oberman to annual meeting of the North American Rail Shippers Association, September 8, 2021, https://www.stb.gov/wp-content/uploads/NARS-Speech-9-8-21.pdf.

216   *"below the bone":* Eleanor Mueller, "The Supply Chain's Little-Known Weakest Link: Railroad Workers," *Politico,* May 16, 2022, https://www.politico

.com/newsletters/weekly-shift/2022/05/16/the-supply-chains-little-known
-weakest-link-railroad-workers-00032624.

216 *"Service is not yet where we want it to be"*: Norfolk Southern earnings call, July 27, 2022, transcript via *The Motley Fool*, https://www.fool.com/earnings/call -transcripts/2022/07/27/norfolk-southern-nsc-q2-2022-earnings-call-transcr/.

216 *"taking advantage of every option"*: Ibid.

217 *"Due to the corporate greed"*: Anthony Gunter resignation letter, August 3, 2022.

219 *$183 million: Report to the President by Emergency Board No. 250,* August 16, 2022, 30, https://nmb.gov/NMB_Application/wp-content/uploads/2022/08/PEB-250 -Report-and-Recommendations.pdf.

219 *"capital investment"*: Ibid., 32.

220 *in half a century*: Peter S. Goodman, "For Rail Workers, Anger Persists over Sick Leave," *New York Times*, October 29, 2022, B1.

220 *"just didn't want to share"*: Ibid.

221 *to destroy organized labor*: Steven Watts, *The People's Tycoon: Henry Ford and the American Century* (New York: Vintage Books, 2006), chapter 22.

221 *"No manufacturer in his right mind"*: Henry Ford, *My Life and Work* (1922; repr., Pantianos Classics, ebook), chapter 8.

222 *"the ruin of all business"*: Richard Snow, *I Invented the Modern Age: The Rise of Henry Ford* (New York: Scribner, 2013), 221.

222 *"A low wage business"*: Ford, *My Life and Work,* chapter 8.

222 *"sociological department"*: Ibid., chapter 11.

222 *a monopolistic grip*: Mark J. Roe, "Dodge v. Ford: What Happened and Why?" European Corporate Governance Institute—Law Working Paper No. 619/2021, November 12, 2021, SSRN, https://ssrn.com/abstract=3943559 or http://dx .doi.org/10.2139/ssrn.3943559.

222 *"Social Justice Animates Ford"*: Watts, *The People's Tycoon,* chapter 10.

222 *massed at Ford plants*: Snow, *I Invented the Modern Age*, 219–20.

222 *handed out marriage licenses:* Ibid., 222.

223 *"Again, we are a family"*: Ibid.

223 *"potential for a strike"*: Interview with Peter Kennedy, October 13, 2022.

223 *"paid sick leave"*: Andrea Hsu, "Some Rail Workers Say Biden 'Turned His Back on Us' in Deal to Avert Rail Strike," NPR, December 2, 2022, https://www.npr .org/2022/12/02/1140265413/rail-workers-biden-unions-freight-railroads -averted-strike.

224 *two bills in the House:* Emily Cochrane, "With Senate Vote, Congress Moves to Avert Rail Strike," *New York Times,* December 2, 2022, A1.

224 *muster the sixty votes:* Ibid.

225 *the company's earnings:* "Welcome to Investor Day," presentation, slide 25, Norfolk Southern, December 6, 2022, http://www.nscorp.com/content/dam /nscorp/get-to-know-ns/investor-relations/Slides/2022_InvestorDay%20 Presentation_Final_For_Web.pdf.

225 *"our quarterly dividend"*: Mark George, in NS WebVideos, "NSC Investor Day 2022 Webcast," YouTube video, December 6, 2022, https://www.youtube.com /watch?v=k8roDn3an5A&t=2087s.

225 *full of hazardous materials:* National Transportation Safety Board press release, February 23, 2023.

225 *Millions of gallons of toxic liquid:* Becky Sullivan, "Here's Why It's Hard to Clean Up Toxic Waste from the East Palestine Train Derailment," NPR, March 3, 2023, https://www.npr.org/2023/03/03/1160481769/east-palestine-derailment -toxic-waste-cleanup.

225 *lobbied the Obama administration:* Julia Rock and Rebecca Burns, "There Will Be More Derailments," *The Lever,* February 10, 2023.

226 *"There's tons of people out there":* Interview with Michael Paul Lindsey II, October 2022.

CHAPTER 15: "SWEATSHOPS ON WHEELS"

229 *three-fourths of all freight:* Andrew W. Hait and Lynda Lee, "What's in That Truck I Just Passed on the Highway?" United States Census Bureau, February 24, 2021, https://www.census.gov/library/stories/2021/02/what-is-in-that-truck -i-just-passed-on-the-highway.html.

230 *eighty thousand more drivers:* Driver Shortage Update 2021, American Trucking Associations, Economics Department, October 25, 2021.

230 *"sort of a warning":* Jim Stinson, "ATA's Costello: Trucking Alone Can't Solve the 80K Driver Shortage," *Transport Dive,* October 25, 2021.

231 *driver turnover rate:* Stephen V. Burks and Kristen Monaco, "Is the U.S. Labor Market for Truck Drivers Broken? An Empirical Analysis Using Nationally Representative Data," IZA Institute of Labor Economics, Discussion Paper No. 11813, September 2018, 31.

231 *A few stayed for years:* Bob Costello and Alan Karickhoff, "Truck Driver Shortage Analysis 2019," American Trucking Associations, July 2019, 4, https://www .trucking.org/sites/default/files/2020-01/ATAs%20Driver%20Shortage%20 Report%202019%20with%20cover.pdf.

232 *more likely to be killed:* US Department of Transportation, *Driving Automation Systems in Long-Haul Trucking and Bus Transit: Preliminary Analysis of Potential Workforce Impacts,* January 2021, https://www.transportation.gov/sites/dot .gov/files/2021-01/Driving%20Automation%20Systems%20in%20Long%20 Haul%20Trucking%20and%20Bus%20Transit%20Preliminary%20Analysis %20of%20Potential%20Workforce%20Impacts.pdf.

233 *"a career of second choice":* Interview with Max Farrell, November 10, 2021.

234 *nearly $12 million:* Alana Semuels, "The Truck Driver Shortage Doesn't Exist. Saying There Is One Makes Conditions Worse for Drivers," *Time,* November 12, 2021.

234 *graduated from such programs:* Ibid.

234 *10 million Americans:* Costello and Karickhoff, "Truck Driver Shortage," 6, n. 10.

235 *"This shortage narrative":* Interview with Steve Viscelli, December 24, 2021.

235 *Trucking firms routinely colluded:* Steve Viscelli, introduction to *The Big Rig: Trucking and the Decline of the American Dream* (Oakland: University of California Press, 2016).

235 *Motor Carrier Act of 1935:* Gary J. Edles, "Motor Carrier Act (1935)," *Encyclopedia*

.com, https://www.encyclopedia.com/history/united-states-and-canada/us
-history/motor-carrier-act.

235 *number of trucking companies:* Viscelli, introduction to *The Big Rig.*
235 *set up centralized terminals:* Ibid.
235 *opportunity for organized labor:* Ibid.
236 *national contract for truck drivers:* Ibid.
236 *represented by a union:* Rachel Premack, "Truck Driver Salaries Have Fallen by as
     Much as 50% Since the 1970s—and Experts Say a Little Known Law Explains
     Why," *Insider,* September 26, 2018.
236 *half again as much:* Viscelli, introduction to *The Big Rig.*
236 *Congress to think tanks:* Ibid.
237 *forces of monopolization:* Ibid.
237 *Ralph Nader's Consumers Union:* Ibid.
237 *Major manufacturers and retailers:* Ibid.
238 *"inflationary government restrictions":* President Jimmy Carter, "Motor Carrier
     Act of 1980 Statement on Signing S. 2245 Into Law," July 1, 1980, UC Santa
     Barbara, The American Presidency Project, https://www.presidency.ucsb.edu
     /documents/motor-carrier-act-1980-statement-signing-s-2245-into-law.
238 *had federal approval:* Viscelli, introduction to *The Big Rig.*
238 *less than 20 percent:* Ibid.
238 *By the year 2000:* Ibid.
238 *more than one-fifth:* Premack, "Truck Driver Salaries."
239 *"sweatshops on wheels":* Michael H. Belzer, *Sweatshops on Wheels: Winners and Los-
     ers in Trucking Deregulation* (New York: Oxford University Press, 2000).
240 *The public health literature:* Adam Hege, Michael K. Lemke, Yorghos Apostolopou-
     los, Sevil Sönmez, "Occupational Health Disparities Among U.S. Long-Haul
     Truck Drivers: The Influence of Work Organization and Sleep on Cardiovascular
     and Metabolic Disease Risk," *PLoS ONE* 13, no. 11 (November 15, 2018), https://
     doi.org/10.1371/journal.pone.0207322.

CHAPTER 16: "THANK YOU FOR WHAT YOU'RE DOING TO KEEP THOSE GROCERY STORE SHELVES STOCKED."

250 *$1.4 billion purchase of Swift:* Tom McGhee, "Swift Sold to Brazilian Firm," *Denver
     Post,* May 29, 2007.
250 *nearly $5 billion:* Chloe Sorvino, *Raw Deal: Hidden Corruption, Corporate Greed,
     and the Fight for the Future of Meat* (New York: Simon & Schuster, 2022), 67–73.
250 *their purchase of Swift:* Ibid., 62–63.
250 *more than $5 billion:* Felipe Marques and James Attwood, "Brazil's Batista Broth-
     ers Are Out of Jail and Worth $6 Billion," *Bloomberg,* July 15, 2021.
250 *turn up for shifts:* Shelly Bradbury, "More Than 800 Greeley Meat Packing Plant
     Workers Call Off as Coronavirus Is Confirmed Among Employees," *Denver Post,*
     March 31, 2020.
252 *five people who died:* US Department of Labor, Occupational Safety and Health
     Administration, "JBS Foods USA Reaches Settlement with OSHA to Develop,
     Implement Infectious Disease Preparedness Plan at Seven Meat Processing

Plants," press release, May 27, 2022, https://www.osha.gov/news/newsreleases/national/05272022.

252 *59,000 meatpacking workers:* US House of Representatives, Select Subcommittee on the Coronavirus Crisis, "Staff Memorandum on Coronavirus Infections and Deaths Among Meatpacking Workers," October 27, 2021, https://docs.house.gov/meetings/VC/VC00/20211027/114179/HHRG-117-VC00-20211027-SD003.pdf.

252 *"all COVID-19 patients":* Email from physician at Moore County Hospital to Cameron Bruett, head of corporate affairs, JBS USA Holdings, April 18, 2020, partially redacted, attached as exhibit to US House of Representatives, Select Subcommittee on the Coronavirus Crisis, "How the Trump Administration Helped the Meatpacking Industry Block Pandemic Worker Protections," May 2022, https://coronavirus-democrats-oversight.house.gov/sites/democrats.coronavirus.house.gov/files/2022.5.12%20-%20SSCC%20report%20Meatpacking%20FINAL.pdf.

253 *"the continued functioning":* The White House, "Executive Order on Delegating Authority Under the DPA with Respect to Food Supply Chain Resources During the National Emergency Caused by the Outbreak of COVID-19," April 28, 2020, https://trumpwhitehouse.archives.gov/presidential-actions/executive-order-delegating-authority-dpa-respect-food-supply-chain-resources-national-emergency-caused-outbreak-covid-19/.

253 *echoed key talking points:* Michael Grabell and Bernice Yeung, "Emails Show the Meatpacking Industry Drafted an Executive Order to Keep Plants Open," *ProPublica,* September 14, 2020.

253 *meeting and corresponding:* Select Subcommittee on the Coronavirus Crisis, "How the Trump Administration Helped."

254 *"perilously close to the edge":* "Smithfield Closes South Dakota Pork Plant Due to Coronavirus," Associated Press, April 12, 2020.

254 *frozen pork:* US Department of Agriculture, National Agricultural Statistics Service, "Cold Storage: March 2020 Highlights," April 22, 2020, 7, https://downloads.usda.library.cornell.edu/usda-esmis/files/pg15bd892/1r66jm331/qb98n127b/cost0420.pdf.

254 *dramatically expanded:* Karl Plume, "Surge in U.S. Pork Exports to China Led by Brazil's JBS, China's WH Group," Reuters, September 22, 2020.

254 *one-third of its output:* "Top 9 Meat Packing plants in the U.S.," IndustrySelect, July 20, 2022, https://www.industryselect.com/blog/the-largest-meatpacking-plants-in-the-us.

254 *"whipped everyone into a frenzy":* Email thread between Sarah Little, vice president of communications, North American Meat Institute, and Robin E. Troy, director of conference services and marketing, April 16, 2020, partially redacted, attached as exhibit to Select Subcommittee on the Coronavirus Crisis, "How the Trump Administration Helped."

254 *"the meat shortage story":* Ibid.

255 *at the US Department of Agriculture:* North American Meat Institute internal emails attached as exhibits to Select Subcommittee on the Coronavirus Crisis, "How the Trump Administration Helped," 14, n. 100.

255   *Vice President Mike Pence:* Ibid., n. 101.

255   *"critical infrastructure":* Ibid., n. 104.

255   *the largest meatpackers:* Ibid., 15.

255   *involve Pence or Trump:* Email from Julie Anna Potts, CEO of North American Meat Institute, to Noel White, CEO Tyson Foods, et al., April 2, 2020, attached as exhibit to Select Subcommittee on the Coronavirus Crisis, "How the Trump Administration Helped," 15–16, n. 111.

255   *"quit your job":* Email from Julie Anna Potts, CEO North American Meat Institute, to USDA Secretary Sonny Perdue, April 3, 2020, attached as exhibit to Select Subcommittee on the Coronavirus Crisis, "How the Trump Administration Helped," 16, n. 112.

255   *"incidents of worker absenteeism":* "Vice President Pence with Coronavirus Task Force Briefing," C-SPAN, video, April 7, 2020, https://www.c-span.org/video/?471020-101/vice-president-pence-coronavirus-task-force-briefing.

255   *"for what you're doing":* Ibid.

256   *$3 billion in dividends:* Brian Deese, Sameera Fazili, and Bharat Ramamurti, "Recent Data Show Dominant Meat Processing Companies Are Taking Advantage of Market Power to Raise Prices and Grow Profit Margins," White House Briefing Room blog, December 10, 2021, https://www.whitehouse.gov/briefing-room/blog/2021/12/10/recent-data-show-dominant-meat-processing-companies-are-taking-advantage-of-market-power-to-raise-prices-and-grow-profit-margins/.

256   *revenues of $22 billion:* JBS SA earnings call with investors, March 25, 2021, transcript via CQ-Roll.

256   *"make us very proud":* Ibid.

257   *fine of $15,615:* US Department of Labor, Occupational Safety and Health Administration, "OSHA News Release—Region 8: U.S. Department of Labor Cites JBS Foods Inc. for Failing to Protect Employees from Exposure to the Coronavirus," press release, September 11, 2020, https://www.osha.gov/news/newsreleases/region8/09112020.

257   *roasts, and other cuts:* US Department of Agriculture data, cited in Peter S. Goodman, "Record Beef Prices, but Ranchers Aren't Cashing In," *New York Times,* December 27, 2021, A1.

CHAPTER 17: **"WE DO NOT HAVE A FREE MARKET."**

259   *record volumes of beef:* Analysis of US Department of Agriculture data furnished by Cassandra Fish.

259   *more than one-fourth:* Average Price: All Uncooked Beef Steaks in U.S. City Average, FRED Economic Data, St. Louis Fed, https://fred.stlouisfed.org/series/APU0000FC3101.

262   *Barry Lynn had once dissected:* Barry C. Lynn, *Cornered: The New Monopoly Capitalism and the Economics of Destruction* (Hoboken, NJ: John Wiley & Sons, 2010), chapter 2.

263   *2000 merger with Nabisco:* Ibid., chapter 3.

263   *43 percent of the market:* Matt Stoller, "Big Bottle: The Baby Formula Nightmare," *BIG,* May 13, 2022.

263   *at the Michigan plant:* Coral Beach, "Former Employee Blows Whistle on Baby Formula Production Plant Tied to Outbreak," *Food Safety News,* April 28, 2022, https://www.foodsafetynews.com/2022/04/former-employee-blows-whistle-on-baby-formula-production-plant-tied-to-outbreak/.

264   *bacterial contamination:* Tom Perkins, "Baby Formula Crisis: Abbott Enriched Shareholders as Factory Needed Repairs, Records Show," *The Guardian,* May 20, 2022.

264   *baby formula in Europe:* Stoller, "Big Bottle."

266   *for baby food, pasta:* Claire Kelloway, "U.S. Food Prices Are Up. Are the Food Corporations to Blame for Taking Advantage?" *Time,* January 14, 2022.

266   *increase in American prices:* Josh Bivens, "Corporate Profits Have Contributed Disproportionately to Inflation. How Should Policy Makers Respond?" *Working Economics Blog,* Economic Policy Institute, April 21, 2022, https://www.epi.org/blog/corporate-profits-have-contributed-disproportionately-to-inflation-how-should-policymakers-respond/.

266   *the highest on record:* Mike Konczal and Niko Lusiani, "Prices, Profits, and Power: An Analysis of 2021 Firm-Level Markups," Roosevelt Institute, June 2022, https://rooseveltinstitute.org/wp-content/uploads/2022/06/RI_PricesProfits Power_202206.pdf.

266   *sweetest profits since 1950:* Matthew Boesler, "Profits Soar as U.S. Corporations Have Best Year Since 1950," *Bloomberg,* March 30, 2022.

266   *"passing along higher cost":* Kroger, Third Quarter 2021 Earnings Call Transcript, Seeking Alpha, December 2, 2021, https://seekingalpha.com/article/4473085-kroger-co-kr-ceo-rodney-mcmullen-on-q3-2021-results-earnings-call-transcript.

267   *"The meat increases we are seeing":* Brian Deese, Sameera Fazili, and Bharat Ramamurti, "Addressing Concentration in the Meat-Processing Industry to Lower Food Prices for American Families," *Briefing Room* (blog), The White House, September 8, 2021, https://www.whitehouse.gov/briefing-room/blog/2021/09/08/addressing-concentration-in-the-meat-processing-industry-to-lower-food-prices-for-american-families/.

268   *outweighed the value:* Christopher Knowlton, *Cattle Kingdom: The Hidden History of the Cowboy West* (New York: Houghton Mifflin Harcourt, 2017), introduction.

268   *Cheyenne, Wyoming:* Ibid.

268   *wealthiest families of the era:* Ibid.

269   *as much as 86 percent:* "Report of the Federal Trade Commission on the Meat-Packing Industry," Summary and Part 1, June 24, 1919, 31, https://www.google.com/books/edition/Report_of_the_Federal_Trade_Commission_o/0GgJAQAAMAAJ?hl=en&gbpv=1.

269   *21,000 head of cattle:* Knowlton, *Cattle Kingdom,* chapter 7.

269   *the national meat supply:* Ibid.

269   *"the blazing midsummer sun":* Upton Sinclair, *The Jungle* (1906; repr., Apple Books, ebook), chapter 26.

269   *"at the mercy":* "Report of the Federal Trade Commission," 24.

270   *"The menace of this concentrated control":* Ibid., 32.

272   *"an end to the arrogance":* "Ronald Reagan's Announcement for Presidential Can-
      didacy, 1979," January 13, 1979, Ronald Reagan Presidential Library & Museum,
      https://www.reaganlibrary.gov/archives/speech/ronald-reagans-announcement
      -presidential-candidacy-1979.

272   *financed by American industry:* Peter S. Goodman, *Davos Man: How the Billionaires
      Devoured the World* (New York: HarperCollins, 2022), 377–79.

272   *were not only antiquated:* Ibid.

273   *Wholesale beef prices did edge down:* US Department of Agriculture, Economic Re-
      search Service, Historical Monthly Price Spread for Beef data, https://www.ers
      .usda.gov/data-products/meat-price-spreads/.

273   *five American slaughterhouses:* Chloe Sorvino, *Raw Deal: Hidden Corruption, Cor-
      porate Greed, and the Fight for the Future of Meat* (New York: Simon & Schuster,
      2022), 26–27.

273   *Fourteen plants had the capacity:* Ibid.

274   *$565 million for Smithfield Beef:* Ibid., 67.

274   *National Beef Packing:* Associated Press, "Brazilian Co. Acquires National Beef,
      Plans to Buy Smithfield Unit to Form No. 1 Processor," *Denver Post,* March 4,
      2008.

275   *"combination of JBS and National":* US Department of Justice, "Justice Depart-
      ment Files Lawsuit to Stop JBS S.A. from Acquiring National Beef Packing Co.,"
      press release, October 20, 2008, https://www.justice.gov/archive/atr/public
      /press_releases/2008/238382.htm.

275   *slashed by 80 percent:* Sorvino, *Raw Deal,* 29.

275   *up from 36 percent:* US Department of Agriculture data, cited in Peter S. Good-
      man, "Record Beef Prices, but Ranchers Aren't Cashing In," *New York Times,* De-
      cember 27, 2021, A1.

275   *of all cattle changed hands:* Interview with Bill Bullard, chief executive officer,
      Ranchers-Cattlemen Action Legal Fund.

276   *all-time low of 37 cents:* USDA data; Goodman, "Record Beef Prices."

277   *reduce their purchases:* Third Consolidated Amended Class Action Complaint,
      In re Cattle Antitrust Litigation, US District Court, District of Minnesota,
      Civil No. 19-cv-1222 (JRT/HB), Doc. 313, 47, https://aglaw.psu.edu/wp
      -content/uploads/2021/04/In-Re-Cattle-Antitrust-Litigation-Third-amended
      -complaint-12.28.20.pdf.

277   *"nothing to do with price":* Email from Sarah Little, December 15, 2021.

CHAPTER 18: **"WE JUST NEED SOME DIVERSITY."**

283   *Biden had added provocation:* Zolan Kanno-Youngs and Peter Baker, "Biden
      Pledges to Defend Taiwan if It Faces a Chinese Attack," *New York Times,* May 24,
      2022, A1.

284   *to publicly clarify:* "No Change in U.S. Policy Towards Taiwan, Says White House
      Official," Reuters, May 23, 2022.

285   *"Our great American companies":* Alan Rappeport and Keith Bradsher, "Trump
      Says He Will Raise Existing Tariffs on Chinese Goods to 30%," *New York Times,*
      August 24, 2019, A1.

286 *a process of "de-risking":* Damien Cave, "How 'Decoupling' from China Became 'De-Risking,'" *New York Times,* May 22, 2023, A6.

286 *American companies in China:* China Business Climate Survey Report, American Chamber of Commerce in China, via Arendse Huld, "2023 AmCham China Business Climate Survey—Insights and Analysis," *China Briefing,* March 10, 2023.

286 *anticipating further deterioration:* Ibid.

286 *European companies:* "Nearly One in Four European Firms Consider Shifting Out of China," *Bloomberg,* June 20, 2022.

287 *"People were astonished":* Interview with Brandon Daniels, July 22, 2022.

287 *flowing into China:* Jason Douglas and Stella Yifan Xie, "Countries Compete to Lure Manufacturers from China," *Wall Street Journal,* March 23, 2023.

287 *drop by 100 million:* Shannon K. O'Neil, *The Globalization Myth: Why Regions Matter* (New Haven, CT: Yale University Press, 2022), chapter 5.

287 *rare earth minerals:* Ibid.

287 *China's government of genocide:* The White House, "Fact Sheet: New U.S. Government Actions on Forced Labor in Xinjiang," press release, June 24, 2021, https://www.whitehouse.gov/briefing-room/statements-releases/2021/06/24/fact-sheet-new-u-s-government-actions-on-forced-labor-in-xinjiang/.

288 *products linked to Xinjiang:* Ana Swanson, "Companies Brace for Impact of New Forced Labor Law," *New York Times,* June 22, 2022, B1.

288 *vulnerable to boycotts:* Peter S. Goodman, Vivian Wang, and Elizabeth Paton, "Global Brands Find It Hard to Untangle Themselves From Xinjiang Cotton," *New York Times,* April 6, 2021, A1.

288 *first months of the pandemic:* Thomas Baumgartner, Yogesh Malik, and Asutosh Padhi, "Reimagining Industrial Supply Chains," McKinsey & Company, August 11, 2020.

288 *exceeded $690 billion:* Juliana Liu, "US-China Trade Defies Talk of Decoupling to Hit Record High in 2022," CNN, February 8, 2023.

289 *dropped below 17 percent:* Bryce Baschuk, "DHL Digs into Trade Data to Track US-China Decoupling," *Bloomberg,* March 15, 2023.

289 *the highest tariffs applied:* Ibid.

289 *"get companies to come back":* Donald J. Trump, Twitter thread, July 12, 2019, https://x.com/realDonaldTrump/status/1149661694735474688?s=20.

289 *three hundred thousand American jobs:* Mark Zandi, Jesse Rogers, and Maria Cosma, "Trade War Chicken: The Tariffs and the Damage Done," Moody's Analytics, September 2019.

289 *global furniture exports:* Lori Ann LaRocco, "China, 'Factory of the World,' Is Losing More of Its Manufacturing and Export Dominance, Latest Data Shows," CNBC, October 20, 2022.

289 *of global footwear exports:* Ibid.

289 *one-third of the work:* Cheng Ting-Fang and Lauly Li, "Apple to Produce Millions of AirPods in Vietnam Amid Pandemic," *Nikkei Asian Review,* May 8, 2020.

289 *the production of iPads:* Nathan Reiff, "Apple to Move Some MacBook Production to Vietnam from China," *Investopedia,* December 20, 2022.

290 *Samsung shifted most:* "Samsung to Relocate Chinese Display," Reuters, June 19, 2020.

290 *reliant on Chinese factories:* Caroline Freund, Aaditya Mattoo, Alen Mulabdic, and Michele Ruta, "Is U.S. Trade Policy Reshaping Global Supply Chains?" World Bank Group, Policy Research Working Paper 10593, October 2023, https://openknowledge.worldbank.org/entities/publication/4edfe909-2761 -4b03-b8a7-153650da7cf6.

290 *Vietnam's apparel industry:* "Clothing Makers Find It Hard to Break with China's Supply Chain," *Bloomberg,* October 30, 2023.

290 *The fourteen members:* Abigail Dahlman and Mary E. Lovely, "Most IPEF Members Became More Dependent on China for Trade over the Last Decade," Peterson Institute for International Economics, October 25, 2023, https://www .piie.com/research/piie-charts/most-ipef-members-became-more-dependent -china-trade-over-last-decade.

290 *"You still depend on China":* Interview with Brad Setser, October 26, 2023.

291 *"bribes have to be paid":* Peter S. Goodman, "China Ventures Southward," *Washington Post,* December 6, 2005.

292 *Between 2014 and 2022:* LaRocco, "China, 'Factory of the World.'"

CHAPTER 19: **"GLOBALIZATION IS ALMOST DEAD."**

298 *"This ain't gray sweat suits":* Jay-Z, "F.U.T.W.," on *Magna Carta . . . Holy Grail* (Roc Nation / Roc-A-Fella Records, 2013).

299 *global greenhouse gas emissions:* Willy C. Shih, "Climate Regulations Are About to Disrupt Global Shipping," *Harvard Business Review,* October 21, 2022, https:// hbr.org/2022/10/climate-regulations-are-about-to-disrupt-global-shipping.

299 *reaching $1.5 trillion:* Ibid.

300 *low-wage countries in Asia:* Lori Ann LaRocco, "China, 'Factory of the World,' Is Losing More of Its Manufacturing and Export Dominance, Latest Data Shows," CNBC, October 20, 2022.

300 *Operation Warp Speed:* Chad P. Bown and Thomas J. Bollyky, "Here's How to Get Billions of COVID-19 Vaccine Doses to the World," *Trade and Investment Policy Watch* (blog), Peterson Institute for International Economics, March 18, 2021, https://www.piie.com/blogs/trade-and-investment-policy-watch/heres-how -get-billions-covid-19-vaccine-doses-world.

300 *Johnson & Johnson:* Ibid.

300 *key suppliers like Corning:* Ibid.

300 *Defense Production Act:* Sharon LaFraniere and Katie Thomas, "Pfizer Nears Deal with Trump Administration to Provide More Vaccine Doses," *New York Times,* December 23, 2020, A1.

301 *expanded on those efforts:* Sharon LaFraniere, "Biden Got the Vaccine Rollout Humming, with Trump's Help," *New York Times,* March 11, 2021, A10.

301 *$52 billion worth:* Zolan Kanno-Youngs, "Biden Signs Industrial Policy Bill Aimed at Bolstering Competition with China," *New York Times,* August 10, 2022, A15.

301 *expanding their factories in China:* Demetri Sevastopulo, "Chipmakers Receiving US Federal Funds Barred from Expanding in China for 10 Years," *Financial Times,* February 28, 2023, https://www.ft.com/content/9f4f9684-088c-45eb -b6bf-962061bfea7b.

301   *"the Chinese Communist Party":* Kanno-Youngs, "Biden Signs."

301   *buying back their shares:* David Shepardson, "U.S. Says It Will Limit Size of Semiconductor Chip Grants," Reuters, July 29, 2022.

302   *the Commerce Department:* Ana Swanson, "Biden Administration Clamps Down on China's Access to Chip Technology," *New York Times,* October 8, 2022, A1.

302   *companies anywhere on earth:* Ibid.

302   *pressed key allies:* Ana Swanson, "Netherlands and Japan Said to Join U.S. in Curbing Chip Technology Sent to China," *New York Times,* January 29, 2022, A12.

302   *spending and tax credits:* Jim Tankersley, "Biden Signs Expansive Health, Climate and Tax Law," *New York Times,* August 16, 2022.

302   *federal tax credits:* Jack Ewing, "Tax Credits for Electric Vehicles Are About to Get Confusing," *New York Times,* December 30, 2022, B1.

303   *40 percent of the value:* Ibid.

303   *$2.8 billion in grants:* Emma Newburger, "Biden Awards $2.8 Billion for Projects to Boost Electric Vehicle Battery Manufacturing," CNBC, October 19, 2022.

303   *Biden administration's embrace:* See, for example, Adam Posen, "America's Zero-Sum Economics Doesn't Add Up," *Foreign Policy* (Spring 2023).

303   *one-fourth of its supply:* Olivia White, Jonathan Woetzel, Sven Smit, et al., "The Complication of Concentration in Global Trade," McKinsey Global Institute, January 12, 2023.

304   *transitioning to cleaner energy:* Michael R. Davidson et al., "Risks of Decoupling from China on Low-Carbon Technologies," *Science* 377, no. 1266 (2022).

304   *the globe to replenish:* See, for example, Pinelopi K. Goldberg and Tristan Reed, "Is the Global Economy Deglobalizing? And If So, Why? And What Is Next?" National Bureau of Economic Research, Working Paper No. 31115, April 2023.

305   *two chip plants in Ohio:* Catherine Thorbecke, "Intel Investing $20 Billion to Bring Chip Manufacturing to Ohio Amid Global Shortage," ABC News, January 21, 2022, https://abcnews.go.com/Business/intel-investing-20-billion-bring-chip-manufacturing-ohio/story?id=82395975.

305   *another American company:* Steve Lohr, "Micron Pledges Up to $100 Billion for Semiconductor Factory in New York," *New York Times,* October 5, 2022, B3.

305   *Taiwan Semiconductor Manufacturing:* Don Clark and Kellen Browning, "In Phoenix, a Taiwanese Chip Giant Builds a Hedge Against China," *New York Times,* December 7, 2022, B1.

305   *four to five times:* Kevin Xu, "The Cost of Deglobalization," *Noema,* February 23, 2023.

305   *half again as much:* Yang Jie, "TSMC's Arizona Chip Plant, Awaiting Biden Visit, Faces Birthing Pains," *Wall Street Journal,* December 5, 2022.

305   *"Globalization is almost dead":* Kevin Xu, "Globalization Is Dead and No One Is Listening," *Interconnected,* December 12, 2022, https://interconnect.substack.com/p/globalization-is-dead-and-no-one.

305   *expand forty factories:* Robert Casanova, "The CHIPS Act Has Already Sparked $200 Billion in Private Investments for U.S. Semiconductor Production," Semiconductor Industry Association, December 14, 2022, https://www.semiconductors.org/the-chips-act-has-already-sparked-200-billion-in-private-investments-for-u-s-semiconductor-production/.

305   *$73 billion of investments:* Camila Domonoske, "2022 Was a Big Year for EV Battery Plants in the U.S. How Big? $73 Billion Big," NPR, December 30, 2022, https://www.npr.org/2022/12/30/1145844885/2022-ev-battery-plants#:~:text=All%20told%2C%20the%20group%20counted,previous%20record%2C%20set%20in%202021.

305   *Ford was breaking ground:* Olivia Evans, "Ford, SK On Bringing 5,000 Jobs to Kentucky with Massive Electric Vehicle Battery Park," *Louisville Courier-Journal,* December 5, 2022.

307   *imported from China:* Harry Moser's testimony before the US-China Economic Security Review Commission, June 9, 2022.

307   *jobs in the pipeline:* "Reshoring Initiative 2022 Data Report," Reshoring Initiative, https://reshorenow.org/blog/reshoring-initiative-2022-data-report/.

CHAPTER 20: **"OKAY, MEXICO, SAVE ME."**

315   *exceeding half a trillion:* Walmart 2022 Annual Report.

318   *group exceeding 40 million:* Sandy Dietrich and Erik Hernandez, "Nearly 68 Million People Spoke a Language Other Than English at Home in 2019," United States Census Bureau, December 6, 2002, https://www.census.gov/library/stories/2022/12/languages-we-speak-in-united-states.html.

320   *he derided as "rapists":* Alexander Burns, "Choice Words from Donald Trump, Presidential Candidate," *New York Times,* June 16, 2015.

320   *"worst trade deal ever made":* "Trump: NAFTA Is 'Worst Trade Deal Ever Made,'" *Bloomberg Quicktake,* October 1, 2018.

320   *"Mexico is the solution":* Interview with Shannon K. O'Neil, November 21, 2022.

320   *exports to the United States:* Robert Koopman, William Powers, Zhi Wang, and Shang-Jin Wei, "Give Credit Where Credit Is Due: Tracing Value Added in Global Production Chains," National Bureau of Economic Research, Working Paper No. 16426, September 2010, Appendix A, 7–8, http://www.nber.org/papers/w16426.

321   *worth nearly $780 billion:* United States Census Bureau, "Trade in Goods with Mexico," https://www.census.gov/foreign-trade/balance/c2010.html.

321   *the United States and China:* United States Census Bureau, "Trade in Goods with China," https://www.census.gov/foreign-trade/balance/c5700.html.

321   *The city of Tijuana:* Shannon K. O'Neil, *The Globalization Myth: Why Regions Matter* (New Haven: Yale University Press, 2022), chapter 4.

322   *$800 million worth every day:* Peter S. Goodman, "How a Texas Border City Is Shaping the Future of Global Trade," *New York Times,* January 9, 2023, B1.

322   *through the local port:* Ibid.

323   *warehouse space:* Data from Prologis.

324   *Tesla, the electric car maker:* Carolina Gonzalez and Maya Averbuch, "Tesla to Build Northern Mexico EV Plant, AMLO Says After Musk Call," *Bloomberg,* February 28, 2023.

324   *$7 billion in foreign investment:* Mexican Ministry of Economy data.

325   *less than $1,500:* Greg Miller, "Container Shipping's 'Big Unwind': Spot Rates Near Pre-COVID Levels," *FreightWaves,* December 27, 2022.

325 *process of deglobalization:* See, for example, Pinelopi K. Goldberg and Tristan Reed, "Is the Global Economy Deglobalizing? And If So, Why? And What Is Next?" Brookings Papers on Economic Activity, March 29, 2023, https://www.brookings.edu/articles/is-the-global-economy-deglobalizing-and-if-so-why-and-what-is-next/.

326 *new configuration taking shape:* For two excellent treatments of this development, see Rana Foroohar, *Homecoming: The Path to Prosperity in a Post-global World* (New York: Crown, 2022), and O'Neil, *The Globalization Myth*.

327 *Only 7 percent:* Peter S. Goodman, "How Columbia Sportswear Is Loosening Its Ties to Asia," *New York Times,* October 26, 2023, B1.

328 *To establish its Hungary plant:* Peter S. Goodman, "Chinese TV Maker Sharpens Focus on Europe," *Washington Post,* December 13, 2004.

CHAPTER 21: **"PEOPLE DON'T WANT TO DO THOSE JOBS."**

338 *nearly half a million:* Aaron Klein, "Not All Robots Take Your Job, Some Become Your Co-worker," *Brookings,* October 30, 2019.

338 *branches with fewer tellers:* James Bessen, "Scarce Skills, Not Scarce Jobs," *The Atlantic,* April 27, 2015, https://www.theatlantic.com/business/archive/2015/04/scarce-skills-not-scarce-jobs/390789/.

339 *In countries like Sweden:* Peter S. Goodman, "The Robots Are Coming, and Sweden Is Fine," *New York Times,* December 28, 2017, A1.

341 *an Amazon warehouse:* Karen Weise and Noam Scheiber, "Amazon Workers on Staten Island Vote to Unionize in Landmark Win for Labor," *New York Times,* April 2, 2022, A1.

341 *"the labor movement":* Jodi Kantor and Karen Weise, "How Two Best Friends Beat Amazon," *New York Times,* April 3, 2022, A1.

CONCLUSION: **"A GREAT SACRIFICE FOR YOU"**

342 *Shipping rates had plunged:* Laura Curtis, "Container Rates Sink Near $1,000 with Demand 'Crashing,'" *Bloomberg,* March 2, 2023.

342 *down 43 percent:* Greg Miller, "West Coast Wipeout: Los Angeles, Long Beach Imports Still Sinking," *American Shipper,* March 17, 2023, https://www.freightwaves.com/news/west-coast-wipeout-los-angeles-long-beach-imports-still-plunging.

342 *"The decline was indeed steep":* Ibid.

343 *limit greenhouse gases:* Willy C. Shih, "Climate Regulations Are About to Disrupt Global Shipping," *Harvard Business Review,* October 21, 2022, https://hbr.org/2022/10/climate-regulations-are-about-to-disrupt-global-shipping.

343 *Corporate Sustainability Reporting Directive:* Kolja Stehl, Leonard Ng, Matt Feehily, and Sidley Austin, "EU Corporate Sustainability Reporting Directive—What Do Companies Need to Know," Harvard Law School Forum on Corporate Governance, August 23, 2022, https://corpgov.law.harvard.edu/2022/08/23

/eu-corporate-sustainability-reporting-directive-what-do-companies-need-to
-know/.

344 *The Panama Canal:* Ben Tracy and Analisa Novak, "Drought Affecting Panama Canal Threatens 40% of World's Cargo Ship Traffic," CBS News, August 23, 2023, https://www.cbsnews.com/news/panama-canal-drought-threatens-global -cargo-ship-traffic/.

344 *its way through Congress:* Laura Curtis, "The Backbone of Global Trade Faces Antitrust Questions in US Congress," *Bloomberg,* March 28, 2023.

345 *probing price fixing:* Hearing at US House of Representatives, Committee on Agriculture, "An Examination of Price Discrepancies, Transparency, and Alleged Unfair Practices in Cattle Markets," April 27, 2022, https://www.youtube.com /watch?v=kSqyELfmMUA.

345 *the bully pulpit:* Nandita Bose, "White House Renews Pressure on Railroads over Paid Sick Leave," Reuters, February 9, 2023.

345 *had struck agreements:* "Union Pacific Railroad Reaches Paid Sick Leave Agreements with Eight Labor Unions," company press release, March 22, 2023, https://www.up.com/media/releases/paid-sick-leave-nr-230322.htm.

345 *"oversimplified market efficiency":* The White House, "Remarks by National Security Advisor Jake Sullivan on Renewing American Economic Leadership at the Brookings Institution," April 27, 2023, https://www.whitehouse.gov /briefing-room/speeches-remarks/2023/04/27/remarks-by-national-security -advisor-jake-sullivan-on-renewing-american-economic-leadership-at-the -brookings-institution/.

346 *eager for a trade deal:* Peter S. Goodman, "How Geopolitics Is Complicating the Move to Clean Energy," *New York Times,* August 22, 2023, B1.

347 *10 percent more:* Laura Alfaro and Davin Chor, "Global Supply Chains: The Looming 'Great Reallocation,'" paper prepared for Jackson Hole Symposium organized by Federal Reserve Bank of Kansas City, August 2023, https://www .kansascityfed.org/documents/9747/JH_Paper_Alfaro.pdf.

348 *moving back to China:* "Clothing Makers Find It Hard to Break with China's Supply Chain," *Bloomberg.*

348 *"works for less pay":* Barbara Ehrenreich, *Nickel and Dimed: On (Not) Getting By in America* (New York: Metropolitan Books, 2001), Evaluation.

INDEX